Sustainable Production, Life Cycle Engineering and Management

Series editors

Christoph Herrmann, Braunschweig, Germany
Sami Kara, Sydney, Australia

Modern production enables a high standard of living worldwide through products and services. Global responsibility requires a comprehensive integration of sustainable development fostered by new paradigms, innovative technologies, methods and tools as well as business models. Minimizing material and energy usage, adapting material and energy flows to better fit natural process capacities, and changing consumption behaviour are important aspects of future production. A life cycle perspective and an integrated economic, ecological and social evaluation are essential requirements in management and engineering. This series will focus on the issues and latest developments towards sustainability in production based on life cycle thinking.

More information about this series at http://www.springer.com/series/10615

Marius Winter

Eco-efficiency of Grinding Processes and Systems

Springer

Marius Winter
Institut für Werkzeugmaschinen und
Fertigungstechnik
Technische Universität Braunschweig
Braunschweig
Germany

ISSN 2194-0541 ISSN 2194-055X (electronic)
Sustainable Production, Life Cycle Engineering and Management
ISBN 978-3-319-25203-2 ISBN 978-3-319-25205-6 (eBook)
DOI 10.1007/978-3-319-25205-6

Library of Congress Control Number: 2015955891

Printed on acid-free paper

This Springer imprint is published by SpringerNature
The registered company is Springer International Publishing AG Switzerland

Foreword

Grinding is a machining process featuring the application of a tool with undefined cutting edges for the processing of hard materials, to create complex profiles and to achieve a superior surface finish. Almost every technological product is directly or indirectly connected with the grinding process during its production. However, grinding is also a time, energy and resource demanding process, resulting in high economic and environmental impacts per removed amount of material. In comparison with machining processes with defined cutting edge, these demands result in a lower eco-efficiency of grinding processes and systems. To maintain the technological, economic and environmental competiveness of grinding processes and systems, an approach is needed which helps to identify and evaluate potentials to increase the eco-efficiency.

In this book, Marius Winter defines an integrated assessment approach to describe, model, evaluate and improve the eco-efficiency of existing and new grinding processes and systems. The foundation and novelty of the proposed concept is the combined empirical and physical modelling of technological, economic and environmental impact indicators based on the relevant energy and resource flows of the grinding process and system. The concept includes a single and integrated evaluation and improvement of different grinding process and system scenarios. Furthermore, an application cycle is provided to transparently determine, evaluate and improve existing and new grinding processes and systems. The developed approach promotes a holistic view on the grinding process and system and is designed to support the designer and operator of such processes and systems to identify decisions which meet the internal and external requirements. The developed approach also shows the potential to be transferred to other manufacturing processes.

With this published work as well as with his active role, Marius Winter has strongly contributed to foster the further development of the topic "cutting fluids" and "mineral oil-free production" in Braunschweig and to foster the development of the Joint German-Australian Research Group "Sustainable Manufacturing and Life Cycle Engineering".

Prof. Christoph Herrmann
Technische Universität Braunschweig

Prof. Sami Kara
The University of New South Wales

Acknowledgments

The present monograph has been written in the context of my work as a research engineer at the Chair of Sustainable Manufacturing and Life Cycle Engineering at the Institute of Machine Tools and Production Technology of the Technische Universität Braunschweig. My sincere gratitude goes to Prof. Dr.-Ing. Christoph Herrmann for the provided working environment, the degree of freedom I was given, the support of my research and the critical discussions of my work. Furthermore, I would like to acknowledge the academic contribution and supervision of Prof. Dr. Konrad Wegener, director of the Institute of Machine Tools and Manufacturing of the Swiss Federal Institute of Technology Zurich, and Prof. Dr.-Ing. Prof. h.c. Klaus Dilger, of the Institute of Joining and Welding of the Technische Universität Braunschweig, for enabling me to successfully finalise this monograph.

During my time as a research engineer I had the pleasure to be part of a highly motivated and supportive team, which allowed me to gather knowledge of multiple research topics. I want to express my sincere thanks to all these colleagues of the whole institute for creating a supportive, collaborative and positive atmosphere, allowing also the discussion of non-research-related topics. In particular, I want to thank Dr.-Ing. Ralf Bock and Dr.-Ing. Sebastian Thiede for their constant support, motivation and fruitful discussions. I am grateful (as well) to M.A. Anne-Marie Schlake for her adjuvant proofread of my manuscript and for providing the opportunity to unravel the daily mysteries of the IWF. Furthermore, I want to thank M.Sc. Nadine Madanchi and Dr. Wen Li for the common interest in research topics about cutting fluids, machining processes, sustainable manufacturing, etc. It was a pleasure to discuss with you new ideas for experiments, projects and publications. A special thank goes also to my officemates Dipl.-Ing. (FH) Stefan Andrew for our quibbling and contradicting discussions and to Dipl.-Wirtsch.-Ing. & Ingénieur diplômé Patricia Egede for our pleasant and fruitful conversation about our research work and work-related topics. Representative for the students that I have supervised in their theses works or assistant jobs I would like to thank Björn Berning, Tim Sieker, Janna Wilke, Alexander Leiden, Joachim Schütte and Christian Sowa for their exceptional work and great support. Many thanks also go to all project partners from industry and research for their trustful collaboration and exchange of experiences.

Beyond that I want to express my genuine gratitude to my friends and family—in particular my parents Regina and Friedrich Wilhelm Winter, for their support and freedom throughout my education. Ohne euch wäre diese Monographie nicht möglich gewesen. Infine e soprattutto, voglio esprimere la mia più profonda gratitudine a Gloria De Angelis per il costante e indescrivibile supporto, il riguardo, la pazienza e l'incoraggiamento durante tutte le fasi della mia laurea, grazie mille.

Contents

Symbols and Abbreviations

Upper Case Symbols

A	Physical model (–)
A_{cf}	Surface area of evaporating cutting fluid (mm^2)
$A_{cu,max}$	Maximal uncut chip area (mm^2)
A_{fm}	Used filtration material surface area (m^2)
$A_{fm,total}$	Total used filtration material surface area (m^2)
A_g	Sectional area cutting groove (mm^2)
A_p	Sectional area piled-up material (mm^2)
A_w	Workpiece surface area (mm^2)
B	Empirical model (–)
C	Proportionality constant ($1/\sqrt{mm}$)
C'	Constraint factor (–)
C_i	Constant (–)
C_{cf}	Cutting fluid costs (€)
C_e	Energy costs (€)
C_{fm}	Filtration material costs (€)
C_{gw}	Grinding wheel costs (€)
C_m	Material costs (€)
C_s	System costs (€)
C_{total}	Total costs (€)
C_w	Waste costs (€)
$CO_{2,p,cf}$	Equivalent CO_2 impact to produced cutting fluid (kg CO_2eq./kg)
$CO_{2,p,e}$	Equivalent CO_2 impact to produced electrical energy (kg CO_2eq./kWh)
$CO_{2,p,fm}$	Equivalent CO_2 impact to produced filtration material (kg CO_2eq./piece)
$CO_{2,p,gw}$	Equivalent CO_2 impact to produced grinding wheel (kg CO_2eq./piece)
$CO_{2,w,cf}$	Equivalent CO_2 impact to disposed cutting fluid (kg CO_2eq./kg)

$CO_{2,w,fm}$	Equivalent CO_2 impact to disposed filtration material (kg CO_2eq./m^2)
$CO_{2,w,gw}$	Equivalent CO_2 impact to disposed grinding wheel residues (kg CO_2eq./kg)
$CO_{2,w,s}$	Equivalent CO_2 impact to disposed grinding swarf (kg CO_2eq./kg)
D	Diffusion coefficient of the cutting fluid (m^2/s)
E	Relative absolute error (–)
E_i	Exponent i (–)
Env_{cf}	Cutting fluid-related equivalent CO_2 impact (kg CO_2eq.)
Env_e	Energy-related equivalent CO_2 impact (kg CO_2eq.)
Env_{fm}	Filtration material-related equivalent CO_2 impact (kg CO_2eq.)
Env_{gw}	Grinding wheel-related equivalent CO_2 impact (kg CO_2eq.)
Env_w	Waste-related equivalent CO_2 impact (kg CO_2eq.)
Env_{total}	Overall equivalent CO_2 impact (kg CO_2eq.)
F_0	Bond strength (N)
F_n	Normal forces (N)
F'_n	Specific normal forces (N/mm)
F_t	Tangential force (N)
F'_t	Specific tangential force (N/mm)
G	Reference point (–)
H	Material hardness (MPa)
$H_{b,g}$	Hardness of the bulk material of the grain (MPa)
$H_{b,w}$	Hardness of the bulk material of the workpiece (MPa)
$H_{s,g}$	Hardness of the grain surface (MPa)
$H_{s,w}$	Hardness of the workpiece surface (MPa)
I	Statistical factor (–)
K	Wear coefficient (1/Pa)
K_i	Constant (–)
K_g	Grain concentration (%)
K_c	Abrasive wear coefficient (1/Pa)
$K_{c,g}$	Abrasive wear coefficient of the grain (1/Pa)
$K_{c,max}$	Maximal abrasive wear coefficient (1/Pa)
L	Sampling length (mm)
M_i	Torque of spindle i (Nm)
N_g	Number of grains (1/mm^2)
$N_{g,d}$	Number of grains created by conditioning (1/mm^2)
P	Normal load (N)
P_0	Machine tool idle power (W)
P_{au}	Power demand of the auxiliary units (W)
P_c	Cutting power (W)
P'_c	Specific cutting power (W/mm)
P_{ca}	Compressor power demand for the compressed air generation (W)

P'_{ca}	Specific compressor power demand for the compressed air generation (W/(Nm³/min))
$P_{CFF,cl}$	Cutting fluid filter power demand cooling unit (W)
$P_{CFF,f}$	Cutting fluid filter power demand filtration pump (W)
$P_{CFF,i}$	Cutting fluid filter power demand of pump i (W)
$P_{CFF,idle}$	Cutting fluid filter power demand idle state (W)
$P_{CFF,l}$	Cutting fluid filter power demand lifting pump (W)
$P_{CFF,proc}$	Cutting fluid filter power demand processing state (decentralised) (W)
$P_{CFF,proc,total}$	Cutting fluid filter power demand processing state (central) (W)
$P_{CFF,s}$	Cutting fluid filter power demand supply pump (W)
P_{cu}	Power demand of the control unit (W)
$P_{EAF,f}$	Exhaust air filter power demand of the filtration unit (W)
$P_{EAF,proc}$	Exhaust air filter power demand processing state (decentralised) (W)
$P_{EAF,proc,total}$	Exhaust air filter power demand processing state (central) (W)
$P_{EAF,v}$	Exhaust air filter power demand of the ventilator drive (W)
P_{el}	Electrical power demand (W)
$P_{GM,idle}$	Grinding machine power demand idle state (W)
$P_{GM,proc,d}$	Grinding machine power demand processing state, dressing (W)
$P_{GM,proc,g}$	Grinding machine power demand processing state, grinding (W)
P_{GS}	Grinding system power demand (W)
P_{hu}	Power demand of the hydraulic unit (W)
P_{mech}	Mechanical pump power demand (W)
$P_{s,d}$	Power demand of the dressing spindle (W)
$P_{s,gw}$	Power demand of the grinding wheel spindle (W)
$P_{s,i}$	Power demand of spindle i (W)
$P_{s,w}$	Power demand of the workpiece spindle (W)
P_{scl}	Power demand of the spindle cooling and lubrication unit (W)
Q_{ca}	Compressed air volume flow (Nm³/h)
Q_{cf}	Cutting fluid volume flow (decentralised) (l/min)
$Q_{cf,total}$	Total cutting fluid volume flow (central) (l/min)
Q_{ea}	Exhaust air volume flow (decentralised) (m³/h)
$Q_{ea,total}$	Total exhaust air volume flow (central) (m³/h)
Q_w	Material removal rate (mm³/s)
Q'_w	Specific material removal rate (mm³/(s mm))
R^2	Coefficient of determination (–)
R_a	Arithmetical mean roughness (μm)
R_a^*	Adjusted arithmetical mean roughness (μm)
R_{s,H_2O}	Specific gas constant for water vapour (J/(kg•K))
Sc	Schmidt number (–)
$\bar{T}$	Arithmetic mean of cutting fluid and air temperature (°C)
$\bar{T}_{cf}$	Mean cutting fluid temperature (°C)
ΔT_{cf}	Tolerated fluid temperature rise (°C)

T_{GM}	Machine tool interior temperature (°C)
T_{max}	Maximal grinding contact zone temperature (°C)
U_d	Dressing overlap ratio (–)
V_K	Grain volume (mm^3)
V_t	Tool wear volume (mm^3)
V'_t	Specific tool wear volume (mm^3/mm)
$V_t'^*$	Adjusted specific tool wear volume (mm^3/mm)
V_w	Workpiece material volume removed (mm^3)
V'_w	Specific workpiece material volume removed (mm^3/mm)
W	Wear volume (mm^3)
Y	Output indicator (–)
Y_i	Observed response (–)
$\bar{Y}_i$	Mean observed response (–)
$\hat{Y}_i$	Predicted response (–)

Lower Case Symbols

a	Half grain width (μm)
$a_{1,i}$	Regression coefficient of spindle i (–)
$a_{2,i}$	Regression coefficient of spindle i (–)
$a_{cf,i}$	Regression coefficient of cutting fluid i (–)
a_e	Cutting depth (μm)
a_{ed}	Dressing depth (μm)
a_p	Cutting width (mm)
a_{pd}	Axial dressing overlap (mm)
b	Half grain height (μm)
b_i	Regression coefficient of spindle i (–)
$b_{cf,i}$	Regression coefficient of cutting fluid i (–)
$b_{cu,max}$	Maximal uncut chip width (μm)
b_d	Effective dresser width (mm)
b_s	Grinding wheel width (mm)
c_0	Coefficient (–)
c_1	Coefficient (–)
$c_{cf,i}$	Regression coefficient of cutting fluid i (–)
c_k	Volume-related grain density (1/mm^3)
c_k^*	Surface area-related grain density (1/mm^2)
$c_{o,e}$	Energy-related overhead cost (%)
$c_{o,m}$	Material-related overhead cost (%)
$c_{o,s}$	System-related overhead cost (%)
$c_{o,w}$	Waste-related overhead cost (%)
$c_{p,cf}$	Specific heat capacity of the cutting fluid (J/(kg•K))
d_{eq}	Equivalent diameter (mm)
$d_{eq,cf,e}$	Equivalent diameter of the evaporation fluid area (m)
d_d	Rotational dresser external diameter (mm)

d_{groove}	Groove depth (μm)
d_k	Mean grain diameter (μm)
d_s	Grinding wheel external diameter (mm)
d_w	Workpiece diameter (mm)
e	Cutting efficiency (–)
f	Corrected target value (–)
f_{cf}	Interfacial friction coefficient of the cutting fluid (–)
f_{ad}	Traverse dressing lead (mm)
$f_{n,i}$	Normal force per grain *i* (N/grain)
f_s	Fraction of grits contacted during one pass of the dresser (–)
$f_{t,i}$	Tangential force per grain *i* (N/grain)
f_{U_d}	Conditioning process weighting factor (–)
h'	Height of the piled-up material (μm)
h_{cu}	Chip thickness (μm)
$h_{cu,max}$	Maximal uncut chip height (μm)
h_d	Number of needed dressing strokes (–)
$h_{d,total}$	Total number of possible dressing times (–)
k	Process physics-related constant (–)
k_m	Water transfer coefficient (–)
l	Dimensionless distance between adjacent tracks (–)
l_c	Cutting length (mm)
l_g	Contact length (mm)
l_{gs}	Grain spacing (μm)
l_w	Workpiece length (mm)
$\dot{m}_{cf,a}$	Cutting fluid loss rate due to aerosols (kg/s)
$m_{cf,a}$	Cutting fluid loss due to aerosols (kg)
$\dot{m}_{cf,e}$	Cutting fluid loss rate due to evaporation (kg/s)
$m_{cf,e}$	Cutting fluid loss due to evaporation (kg)
$\dot{m}_{cf,fm}$	Cutting fluid loss rate due to filtration material wetting (kg/kg)
$m_{cf,fm}$	Cutting fluid loss due to filtration material wetting (kg)
$\dot{m}_{cf,l}$	Cutting fluid loss/demand rate (kg/s)
$m_{cf,l}$	Cutting fluid loss/demand (kg)
$\dot{m}_{cf,s}$	Cutting fluid loss rate due to grinding swarf wetting (kg/kg)
$m_{cf,s}$	Cutting fluid loss due to grinding swarf wetting (kg)
$\dot{m}_{cf,w}$	Cutting fluid loss rate due to workpiece wetting (kg/piece)
$m_{cf,w}$	Cutting fluid loss due to workpiece wetting (kg)
n	Exponent (–)
n_i	Rotational speed of spindle *i* (1/min)
n_s	Grinding wheel rotational speed (1/min)
n_{sd}	Grinding wheel rotational speed when dressing (1/min)
n_w	Workpiece rotational speed (1/min)
p_{cf}	Pressure cutting fluid (bar)
p_e	Energy procurement price (€/kWh)
p_l	Labour price (€/h)

p_{m}	Total grinding system machine-hour rate (€/h)
$p_{m,CFF}$	Cutting fluid filter machine-hour rate (€/h)
$p_{m,EAF}$	Exhaust air filter machine-hour rate (€/h)
$p_{m,GM}$	Grinding machine machine-hour rate (€/h)
$p_{p,cf}$	Cutting fluid procurement price (€/kg)
$p_{p,fm}$	Filtration material procurement price (€/m^2)
$p_{p,gw}$	Grinding wheel procurement price (€/piece)
$p_{s,air}$	Water vapour pressure of the ambient air (bar)
p_{s,H_2O}	Saturation pressure of water vapour (bar)
$p_{w,cf}$	Cutting fluid disposal price (€/kg)
$p_{w,fm}$	Filtration material disposal price (€/m^2)
$p_{w,gm}$	Grinding wheel disposal price (€/piece)
$p_{w,s}$	Grinding swarf disposal price (€/kg)
q	Deviation or offset (–)
q_{d}	Dressing speed ratio (–)
q_{m}	Grit aspect ratio (–)
r	Ratio between uncut chip width and height (–)
r_{d}	Radius of the dresser (mm)
r_{mean}	Mean grain radius (μm)
r_{g}	Grain tip radius (μm)
Δr_{s}	Radial grinding wheel wear (μm)
s	Sliding distance (m)
t_{c}	Grinding time (s)
t_{cf}	Cutting fluid dripping time (s)
$t_{cf,l}$	Cutting fluid service life (s)
t_{d}	Dressing time (s)
t_{idle}	Idle time (s)
t_{h}	Handling time (s)
t_{p}	Processing time (s)
t_{t}	Tool travel time (s)
u	Process and system parameter and state conditions (–)
v_{air}	Air velocity (m/s)
v_{c}	Cutting speed (m/s)
v_{cf}	Cutting fluid jet velocity (m/s)
v_{d}	Dressing tool speed (m/s)
v_{fad}	Traverse dressing speed (m/s)
v_{fr}	Radial infeed speed (m/s)
v_{s}	Grinding wheel speed (m/s)
v_{w}	Workpiece speed (m/s)
w_{m}	Average grit mesh size (μm)
w	Target value (–)
x_{i}	Input variables (–)

y	Actual value (–)
z	Disturbance value (–)

Greek Symbols

α	Attack angle (°)
$\beta_{u/b}$	Water transfer coefficient (according to VDI 2089-1) (–)
γ	Angle of piled-up material (°)
δ_t	Grain fracture strength (GPa)
η	Efficiency (–)
η_{CFO}	Cutting fluid orifice efficiency (–)
η_{el}	Pump drive electrical efficiency (–)
η_{sust}	Sustainable efficiency indicator (–)
θ_{cf}	Cutting fluid water content (%)
μ	Friction coefficient/Force ratio (–)
ν	Grain loss factor (–)
ν_{cf}	Kinematic viscosity of the cutting fluid (mm^2/s)
ρ_{cf}	Density of the cutting fluid (kg/m^3)
ρ_{gw}	Density of the grinding wheel (kg/m^3)
ρ_w	Density of the workpiece (kg/m^3)
σ_{cf}	Surface tension of the cutting fluid (mN/m)
τ	Distance between the mean line of an array of grooves and the initial workpiece surface level (μm)
ϕ_{air}	Relative air humidity (%)

Abbreviations

BCSD	Business Council for Sustainable Development
cBN	Cubic boron nitride
CCA	Conventional cost accounting
CNC	Computer numerical controlled
cvx	Disciplined convex programming
DIN	Deutsches Institut für Normung
EEA	European Environment Agency
EE-portfolio	Portfolio of the environmental impact reduction factor and the economic value factor
EN	European Norm
EU	European Union
EU27	European Union (member states until June 2013)
EU28	European Union (member states since June 2013)
FEM	Finite element method
GPI	Green productivity index
GWP	Global warming potential

hBN	Hexagonal boron nitride
HPHT	High pressure and high temperature
HPJAM	High pressure jet-assisted machining
ISO	International Standard Organisation
LCA	Life cycle assessment
LCC	Life cycle costing
MFCA	Material flow cost accounting
MRR	Material removal rate
OECD	Organisation for Economic Co-operation and Development
PROMETHEE	Preference Ranking Organisation Method for Enrichment of Evaluations
SEC	Specific energy consumption
TE-portfolio	Portfolio of the environmental impact reduction factor and the technological value factor
UN	United Nations
VDI	Verein Deutscher Ingenieure
VDMA	Verband Deutscher Maschinen- und Anlagenbau
WBCSD	World Business Council for Sustainable Development

List of Figures

List of Tables

Chapter 1
Introduction

Abstract This chapter explains the motivation and the scope of action as well as the research objective and the work structure.

1.1 Motivation and Scope of Action

The 20th century was characterised by remarkable economic growth, coupled with a dramatic increase in energy and resource demand (Fischer-Kowalski et al. 2011, p. XIII). This increased demand has resulted in economic and environmental burdens caused by the extraction, transportation and transformation of raw materials and energy carriers. However, in the last decades, a growing political and social awareness about environmental issues has spurred demand for more sustainable activities and a reduced environmental impact (Weizsäcker et al. 1997, p. 9; Herrmann 2010, p. 44f.). This impetus led to the ambitious goal of pursuing a "… development that meets the needs of the present without compromising the ability of future generations to meet their own needs" (Brundtland Report 1987, p. 15) and resulted in the definition of the concept of "sustainable development". This concept aims to improve the environmental, economic and social dimensions equally (Herrmann 2010, p. 46), as represented by the three pillars of sustainable development (Fig. 1.1).

However, in its initial stages, the formulation of this concept did not provide any strategies to enable the change from unsustainable to sustainable development (Mickwitz et al. 2005, p. 1603). The first attempt to address the shortfall was Schaltegger and Sturm's (1989) "eco-efficiency" approach, which took into account the interaction between the economy and the environment (Schaltegger and Sturm 1989, p. 15ff). Three years later, Schaltegger and Sturm's idea gained traction at the United Nations Conference on Environment and Development held in Rio de Janeiro whose outcome, in accordance with the eco-efficiency approach, highlighted the need for "… efficiency in production and changes in consumption patterns to emphasize optimization of resource use and minimization of waste" (UN 1992, p. 20). Efficiency

M. Winter, *Eco-efficiency of Grinding Processes and Systems*,
Sustainable Production, Life Cycle Engineering and Management,
DOI 10.1007/978-3-319-25205-6_1

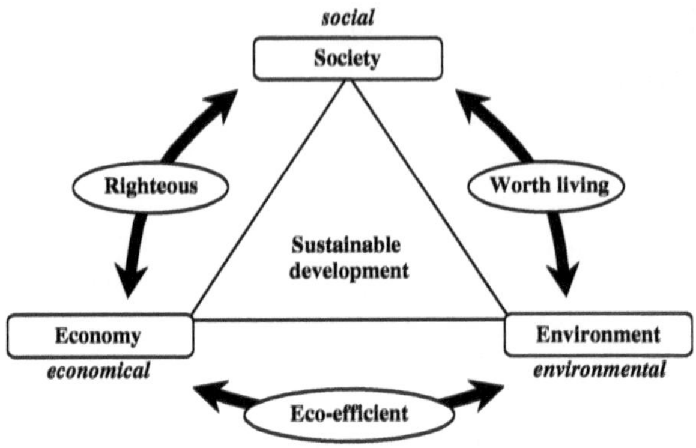

Fig. 1.1 The three pillars of sustainable development [according to (Herrmann 2010, p. 47)]

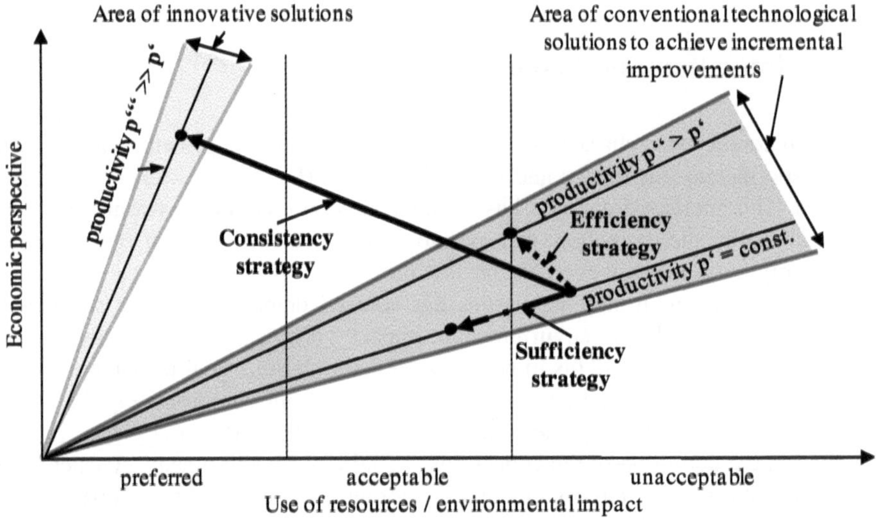

Fig. 1.2 The three strategies of sustainable development [Thiede 2012, p. 3 (based on Schmidt 2007, p. 7)]

focuses on the minimisation of material and energy usage during the production of goods, while the change of consumption patterns (also known as sufficiency) consists of alterations in the demand for energy and resources as a result of a change in user behaviour (Fig. 1.2) (Huber 2000a, p. 275). Efficiency and sufficiency, however, are only viable for a limited time and focus on the improvement of an initial system instead of creating a new one. To overcome this drawback, a third strategy was introduced: consistency. Consistency is based on the reconcilability of economy growth and environmental protection, through the compliance of anthropogenic material and energy flows with natural metabolic products (Herrmann 2010, p. 51). The consistency strategy can achieve significant improvements in sustainability; its implementation, however, is connected with certain efforts in terms of time and cost (Huber 2000b, p. 107ff; Thiede 2012, p. 3f.).

In order to apply the three strategies of sustainable development, it is necessary to identify relevant industrial sectors and processes that have a major influence on the economy, society and the environment. In this context, the manufacturing industry and its connected production processes have a great impact, due to their predominant influence on the environment and on the economy and society (Thiede 2012, p. 2; Pflieger and Schmid 1975, p. 635ff.). The manufacturing industries alone contributed in 2013 with 15.1 % to the gross value added of the EU28 (Eurostat 2015a) and employed circa 14.0 % of the total EU28 labour force (Eurostat 2015b). Within the manufacturing industries, companies focusing on the production of machinery or means of transport rely on machine tools for the machining of materials. According to estimates, the electricity demand of 3.5 million installed machine tools in the EU27 (state 2009) is up to 200–300 TWh per annum (Schischke et al. 2012a, p. 8, 2012b, p. 39). Due to the necessary transformation of primary energy carriers into electrical energy, carbon dioxide emissions of around 25.8 kilotons arise every year (state 2014) (EU 2014, p. 174f.). Furthermore, if the average electricity price of the EU28 industry is assumed [0.1438 €/kWh in 2012 (EU 2014, p. 129)], an economic value of 43.1 milliard € arises.

One of the prominent manufacturing processes in this context is grinding. The production of almost every technological product is connected with the grinding process (Malkin and Gou 2008, p. 1). In 2009, about 28 % of all installed computer numerically controlled (CNC) machine tools (about 750.000 machines in the EU27) were grinding machines (Schischke et al. 2012b, p. 12ff.). Around 20–25 % of the total expenditure for all machining processes in industrial countries stems from grinding processes (Malkin and Gou 2008, p. 1).

Recent investigations on the application of the three sustainable development strategies in the field of manufacturing systems and machine tools indicate the need to take into account energy and resource flows between and within the production processes and systems (Thiede 2012, p. 175ff; Zein 2012, p. 135ff.). Particularly in the case of machining processes with defined cutting edges, investigations focusing on how to increase energy and resource efficiency were already conducted (Kuhrke 2011, p. 1ff; Schlosser 2013, p. 1ff). In the case of grinding processes and systems, potential improvements to eco-efficiency have not yet been explored due to:

- the absence of information on the environmental, economic and technological impact of the necessary material and energy flows,
- the complexity of the interactions within the grinding process and system limiting predictability,
- the difficulty in evaluating and improving the grinding process and system given the various interdependencies.

1.2 Research Objectives and Work Structure

Considering the challenges and requirements of manufacturing companies to reduce the environmental impact while maintaining competitive, this research work has the central scientific objective to develop an integrated assessment approach, which enfolds economic, environmental and technological indicators. Antagonistic effects among the indicators are thus taken into account to conduct a purposive and case-dependent description, modelling, evaluation and improvement of the eco-efficiency of existing and new to plan grinding processes and systems. For this purpose, the present work will consider the interdependencies between the grinding process and system, based on the representation of the energy and material flows in parameter-and property-dependent physical and empirical formulations.

The proposed approach should support designers and operators of such processes and systems in making decisions, which meet the existing internal and external requirements. To this purpose, the following aspects will be considered:

- describing the investigated grinding processes and systems on the basis of parameter and properties,
- modelling the grinding process and system by using empirical and physical formulations taking into account different processes and conditioning parameters, grinding wheel, workpiece, cutting fluid and machinery characteristics,
- evaluating single and integrated impact depending on specific grinding process and system conditions,
- improving environmental, economic and technological indicators to achieve an eco-efficient grinding process and system.

The approach needs to consider the challenges faced in complicated decision-making situations. The application shall foster a deeper understanding of the relationship between the economic, environmental and technological aspects of the grinding process and system. Furthermore, process parameters and system configurations shall be derived resulting in an increased eco-efficiency. The present investigation is structured into seven chapters, and the research method is illustrated in Fig. 1.3.

Following this introduction, the second chapter presents the theoretical foundations of this work. This includes the definition and characterisation of the grinding process and system, as well as modelling and eco-efficiency evaluation. The definition of grinding process and system focuses on the elements and relationships involved in chip formation, grinding process and system. The paragraph on modelling explores the different modelling types, approaches and indicators. The next subsection deals with the origin, definition and assessment of eco-efficiency. The chapter closes with a preliminary conclusion.

The third chapter presents a review and comparative evaluation of research approaches in the evaluation of the eco-efficiency of processes and systems. At the core of the review is the clustering of the identified research approaches according to investigation level and scope. The clustered approaches are then presented in a

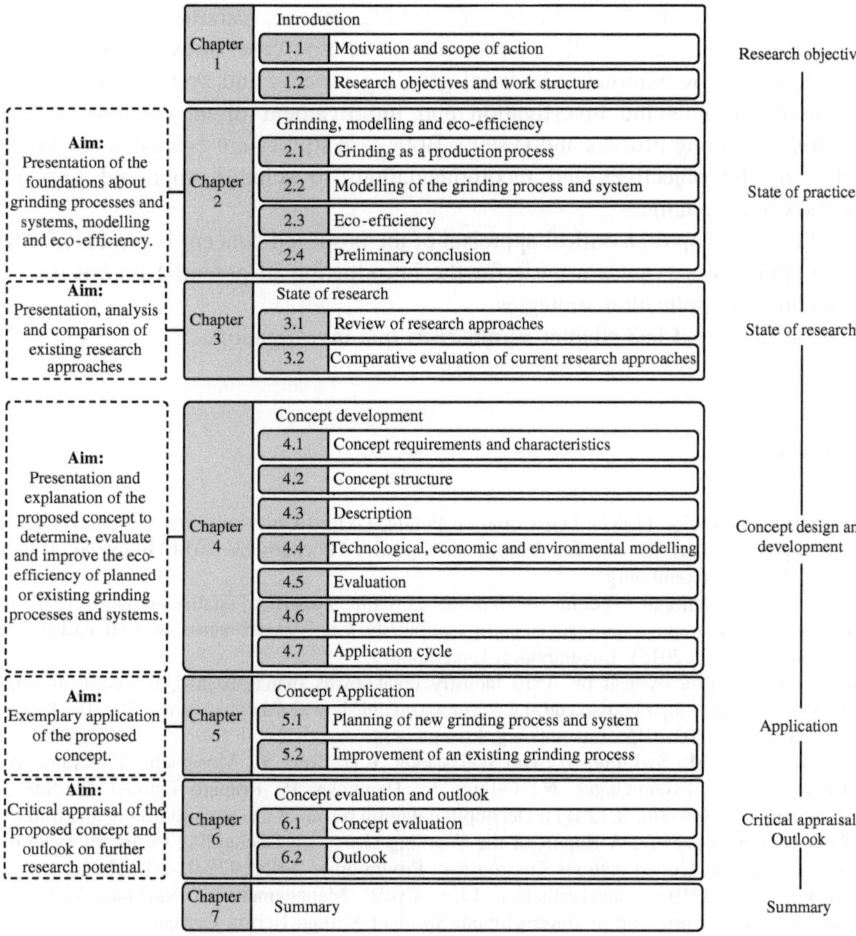

Fig. 1.3 Structure of the research approach

detailed analysis. Research approaches exploring the aforementioned research objective are a prominent part of the comparative evaluation. The selected approaches are compared based on prior defined comparison criteria. Based on the present work, the case is made for further research exploring this topic.

Based on the research question, the fourth chapter proposes a concept to determine the eco-efficiency of grinding processes and systems. The chapter starts with the presentation of the concept requirements and characteristics followed by its structure. The elements' description, modelling, evaluation and improvement of the concept are then presented in detail. The chapter closes with the derivation of an application cycle to determine the eco-efficiency of grinding processes and systems.

The fifth chapter comprises an exemplary application of the proposed concept using the application cycle presented in chapter four. Two case studies are used to analyse and evaluate the influence of varying process parameters and system configurations on the overall eco-efficiency. The first case study focuses on the planning of a new external cylindrical grinding process and system. The second case study presents the investigation and improvement of an existing internal cylindrical grinding process and system. Both case studies are evaluated according to the specific objectives and recommendations to achieve improved grinding processes and systems.

In the sixth chapter, a critical appraisal of the proposed concept is presented. The critical appraisal serves as a basis for the introduction of potential improvements and additional application strategies.

The seventh and last chapter summarises this investigation.

References

Brundtland Commission (1987): Our Common Future. Oxford University Press, Oxford, UK.

EU (European Union) (2014): EU Energy in Figures - Pocketbook 2014. European Union, Luxembourg, Luxembourg.

Eurostat (2015a): National Accounts by 10 branches [nama_nace10_c] (statistical data). Available online: http://appsso.eurostat.ec.europa.eu/nui/show.do?dataset=nama_nace10_c&lang=en (last access 20.03.2015), Luxembourg, Luxembourg.

Eurostat (2015b): Employment by A*10 industry breakdowns [nama_10_a10_e] (statistical data). Available online: http://appsso.eurostat.ec.europa.eu/nui/show.do?dataset=nama_10_a10_e&lang=en (last access 20.03.2015), Luxembourg, Luxembourg.

Fischer-Kowalski, M., Swilling, M., von Weizsäcker, E.U., Ren, Y., Moriguchi, Y., Crane, W., Krausmann, F., Eisenmenger, N., Giljum, S., Hennicke, P., Romero Lankao, P., Siriban Manalang, A., Sewerin, S. (2011): Decoupling natural resource use and environmental impacts from economic growth, A Report of the Working Group on Decoupling to the International Resource Panel. United Nations Environment Programme, ISBN: 978-92-807-3167-5.

Herrmann, C. (2010): Ganzheitliches Life Cycle Management - Nachhaltigkeit und Lebenszyklusorientierung in Unternehmen. Springer Verlag, Berlin, Germany.

Huber, J. (2000a): Towards Industrial Ecology. Sustainable Development as a Concept of Ecological Modernization. Journal of Environmental Policy and Planning, 2/4:269–285.

Huber, J. (2000b): Industrielle Ökologie. Konsistenz, Effizienz und Suffizienz in zyklusanalytischer Betrachtung. In: Kreibich, R., Simonis, U. (Ed.) (2000): Global Change - Globaler Wandel. Ursachenkomplexe und Lösungsansätze, 107–124, Baden-Baden, Germany.

Kuhrke, B. (2011): Methode zur Energie- und Medienbedarfsbewertung spanender Werkzeugmaschinen. Dr.-Ing. Dissertation, Technische Universität Darmstadt, epubli GmbH, Berlin, Germany.

Malkin, S., Guo, C. (2008): Grinding Technology: Theory and Applications of Machining with Abrasives. Industrial Press, New York, USA.

Mickwitz, P., Melanen, M., Rosenström, U., Seppälä, J. (2005): Regional eco-efficiency indicators - a participatory approach. Journal of Cleaner Production, 14:1603–1611.

Pflieger, G.F., Schmid, J. (1975): Bedeutung der Fertigungstechnik für die Industrialisierung von Entwicklungsländern. WT. Werkstattstechnik, 65/10:635–638.

Schaltegger, S., Sturm, A. (1989): Ökologieinduzierte Entscheidungsprobleme des Managements: Ansatzpunkte zur Ausgestaltung von Instrumenten. WWZ-Discussion Paper No. 8914.

Schischke, K., Hohwieler, E., Feitscher, R., König, J., Kreuschner, S., Wilpert, P., Nissen, N.F. (2012a): Energy-Using Product Group Analysis - Lot 5, Machine tools and related machinery, Task 1 Report – Definition, Berlin, Germany.

Schischke, K., Hohwieler, E., Feitscher, R., König, J., Kreuschner, S., Wilpert, P., Nissen, N.F. (2012b): Energy-Using Product Group Analysis - Lot 5, Machine tools and related machinery, Task 2 Report – Economic and Market Analysis, Berlin, Germany.

Schlosser, R. (2013): Methodik zur Prognose der Nachhaltigkeit des Energie- und Stoffeinsatzes spanender Fertigungsprozesse. Dr.-Ing. Dissertation, RWTH Aachen, Shaker Verlag, Aachen, Germany.

Schmidt, M. (2007): Rohstoffe und Ressourceneffizienz - bereit für den Wettbewerb um Nachhaltigkeit (Presentation). Statussemeinar 2007, Berlin, Germany.

Thiede, S. (2012): Energy Efficiency in Manufacturing Systems. Dr.-Ing. Dissertation, Technische Universität Braunschweig, Springer Verlag, Berlin, Germany.

UN (United Nations) (Ed.) (1992): Earth Summit. Agenda 21. The United Nations Programme of Action from Rio. The final text of agreements negotiated by Governments at the United Nations Conference on Environment and Development (UNCED). UN, Rio de Janerio, Brazil.

Weizsäcker, E.U. von, Lovins, A.B., Lovins, L.H. (1997): Faktor Vier: Doppelter Wohlstand - halbierter Naturverbrauch. Der neue Bericht an den Club of Rome. Droemer Knaur, München, Germany.

Zein, A. (2012): Transition Towards Energy Efficient Machine Tools. Dr.-Ing. Dissertation, Technische Universität Braunschweig, Springer Verlag, Berlin, Germany.

Chapter 2
Grinding, Modelling and Eco-efficiency

Abstract This chapter presents the theoretical foundations of this work. Firstly, the basics of the grinding process and the elements of the grinding system are illustrated. Secondly, modelling approaches and indicators are described. Lastly, the origin, definitions and approaches on how to achieve eco-efficiency are presented. The chapter ends with a preliminary conclusion.

2.1 Grinding as a Production Process

DIN 8589 defines the term grinding, as machining with a tool whose number of cutting edges, rake angle geometries and position of cutting edges to the workpiece surface is not defined (DIN 8589-0 2003, p. 4). Therefore, the term describes the application of track-bound hard materials, which differ slightly in their irregular shape and in their relative position to each other, for the machining of materials.

In industrial countries, the production process grinding accounts for 20–25 % of the total expenditure of all machining processes (Malkin and Gou 2008, p. 1). In 2009, 28 % of all the installed CNC machine tools (about 750.000) in the European Union (EU27) were grinding machines (Schischke et al. 2012c, p. 12ff.). Grinding is applied either to attain high material removal rates or to achieve defined surface qualities, accuracy and certain workpiece characteristics. In both cases, it is commonly applied for the machining of hard-to-cut materials (Rowe 2009, p. 2f.; Marinescu et al. 2007, p. 3ff.), such as hardened steel, aerospace alloys, ceramics and cemented carbide (Rowe 2009, p. 2).

The production of almost every technological product is either directly or indirectly connected with the grinding process (Malkin and Gou 2008, p. 1). The direct connection arises as a result of the grinding process being applied at certain stages of the manufacturing process. Typically, grinding is one of the last manufacturing steps in the production chain. The process can be used to achieve high accuracy while machining hard materials. The machined parts are very valuable, and defects due to grinding can lead to high costs (Rowe 2009, p. 4). The indirect

M. Winter, *Eco-efficiency of Grinding Processes and Systems*,
Sustainable Production, Life Cycle Engineering and Management,
DOI 10.1007/978-3-319-25205-6_2

connection arises due to the impact of the grinding process on the accuracy of other machining processes, such as milling, turning or drilling, owing to the application of the grinding process in the production of components (e.g. bearings, guides, gears) as well as tools (e.g. drills, cutting inserts) for these machine tools.

Nonetheless, grinding is in competition with other machining processes, such as turning or milling. When it comes to achieving high volumes of removed material, the grinding process has lower economic and technological productivity than the milling and turning processes. However, due to the emergence of near-net shape production techniques (for example precise casting and forging), the need to remove high volumes of material decreases and grinding becomes again a viable option to increase process productivity by achieving the desired geometry in one machining process without prior turning or milling processes (Malkin and Gou 2008, p. 3).

Grinding is a key technology to meet predefined technological requirements. However, its application needs to be considered in connection with economic and environmental boundaries set by customers, competitors, public institutions, etc. To achieve a high-quality grinding process combined with a low economic and environmental impact, knowledge about the process as well as the necessary demand of energy and resources is needed.

Levels of Grinding

Grinding is performed within a system consisting of different subsystems and components. In accordance with DIN 25424-1, the following terms can be identified (DIN 25424-1 1990, p. 2):

- **System (overall)**: "an aggregation of technical-organisational means for the autonomous realisation of a complex of tasks" (e.g. grinding machine),
- **Subsystem**: "a combination of components, to solve an associated task within a technical system" (e.g. turret wheel head),
- **Component**: "the smallest item under consideration of a technical system. One or more functional elements are related to each component" (e.g. spindle drive) and
- **Functional element**: "the smallest item under consideration in a functional system. It can only describe one elemental function" (e.g. commutator ring inside the spindle).

By using the system theory, relationships between these terms can be described. Ropohl defines three system concepts: the functional, the structural and the hierarchical (Ropohl 2009, p. 75). These concepts are convertible into each other: functional purposes, for example, can be described using a structure that is part of multiple hierarchical levels. Therefore, the functions define the structure and the hierarchy of the system (Ropohl 2009, p. 77).

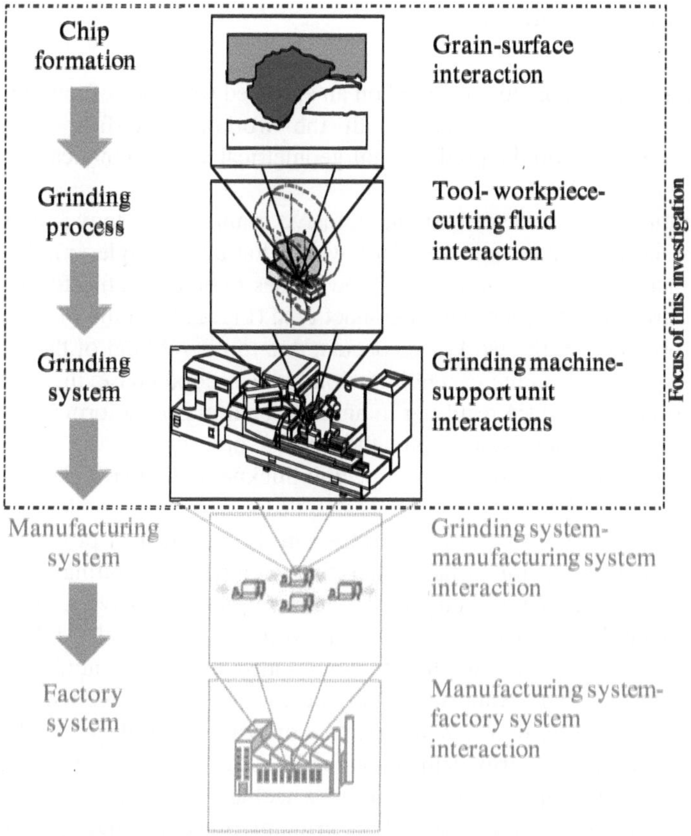

Fig. 2.1 Hierarchical levels in grinding

Different hierarchical levels and corresponding systems can be found in grinding. Each of these systems can be expressed as a subsystem, consisting of components and functional elements of another system (Fig. 2.1).

The lowest system level is chip formation, characterised by the grain—surface interaction. Chip formation is part of the next level, the grinding process, consisting of the tool–workpiece–cutting fluid interaction. This system is incorporated into the grinding system, including a grinding machine–support unit interaction. The next, higher level is the manufacturing system, which represents a subsystem of the factory system. Both structural system concept and functional concept can be used to describe the independency of different systems if these are unknown. In the next sections, chip formation, grinding process and grinding system are described in detail.

2.1.1 Chip Formation

The material removal process is based on an enforced sequence of randomly located single cutting edge engagements with the workpiece surface material. The engagement depends on the predominant geometrical and kinematical relationships (Werner 1971, p. 12).

The cutting edge is located on the abrasive grain, which moves along a fixed grain path and penetrates the workpiece with a flat attack angle (α). The abrasive grain interacts with the workpiece in three phases to achieve chip formation. Each phase is characterised by a different contact type (Fig. 2.2). In the first phase (I), an elastic deformation of the workpiece occurs, due to the rubbing of the grain on the workpiece surface. With increasing intrusion in the second phase, the grain ploughs into the workpiece surface resulting in an additional plastic deformation (II). After exceeding a critical intrusion depth, the cutting process and the chip formation occurs (III), leading to a chip with a specific thickness (h_{cu}) (Klocke 2009, p. 8/9; Rowe 2009, p. 341ff.).

During chip formation, energy is needed in the mechanical processes (cutting, ploughing and rubbing) for surface generation. Compared to cutting processes with geometrically defined cutting edges, the abrasive processes have a higher specific energy demand for the removal of one unit volume of material (Malkin and Guo 2008, p. 120). This is due to the aforementioned undefined number, shape and position of cutting edges. Almost all of the specific energy is converted into heat, and only a minor share is used to form the material or to accelerate the chips (Rowe 2009, p. 370). The degree of energy demand depends on the predominant geometrical and kinematical relationships between grinding wheel and workpiece, as well as on the physical and chemical composition of the grinding process (grinding wheel, cutting fluid and workpiece). These components are described in the next section.

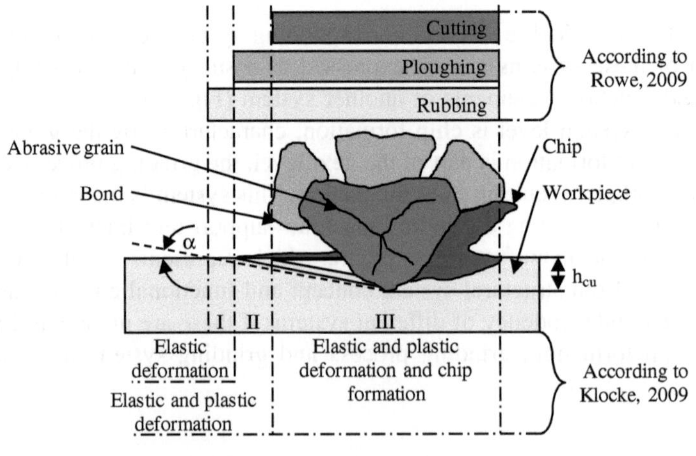

Fig. 2.2 Phases of chip formation and underlying contact types [(Linke and Dornfeld 2012, p. 415) (based on Klocke 2009, p. 8; Rowe 2009, p. 341ff.)]

2.1.2 Grinding Process

The predominant geometrical and kinematical relationships and the related parameters are presented in Fig. 2.3 for the following orthogonal grinding process types: internal cylindrical grinding, external cylindrical grinding and surface grinding.

The input parameters of the grinding process result from the following geometrical and kinematical relationships: cutting depth (a_e), cutting speed (v_c) and workpiece speed (v_w). In the case of internal and external cylindrical grinding processes, the reference point (G), located on the grinding wheel, moves nominally on a cutting depth-dependent arithmetic spiral into the workpiece (if the workpiece speed is constant).

The workpiece speed (v_w) (Eq. 2.1) and grinding wheel speed (v_s) (Eq. 2.2) are calculated based on the rotational speed of the workpiece (n_w) and grinding wheel (n_s) at the workpiece diameter (d_w) and grinding wheel external diameter (d_s), respectively. Equations (2.1) and (2.2) follow the approximation that the grinding wheel speed (v_s) equals the cutting speed (v_c) (Böge 2007, p. 368).

$$v_w = \pi \cdot n_w \cdot d_w \tag{2.1}$$

$$v_s = \pi \cdot n_s \cdot d_s = v_c \tag{2.2}$$

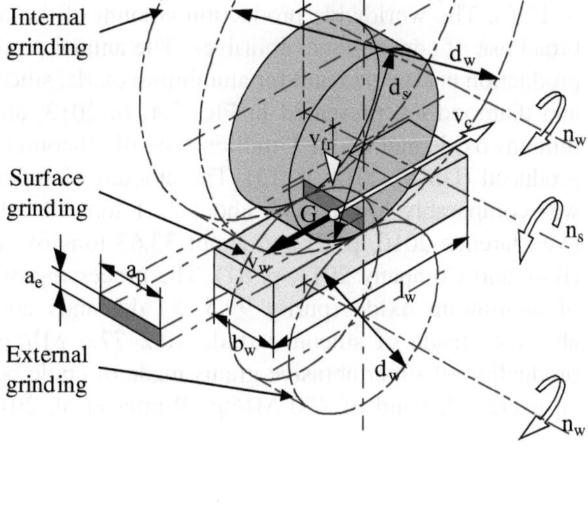

Fig. 2.3 Geometrical and kinematical relationships in orthogonal grinding processes (based on Saljé and Möhlen 1986, p. 249)

An important key figure in describing the grinding process is the material removal rate (Q_w) (Eq. 2.3) and the specific material removal rate (Q'_w) (Eq. 2.4). Both values describe the material removed per second; however, the latter also describes the amount related to 1 mm cutting width (a_p).

$$Q_w = v_w \cdot a_e \cdot a_p \tag{2.3}$$

$$Q'_w = \frac{Q_w}{a_p} = v_w \cdot a_e \tag{2.4}$$

2.1.2.1 Grinding Wheel

The structure and composition of the grinding wheel is dependent upon the material to be machined, the necessary material removal rate as well as the required workpiece accuracy and surface quality. It is commonly composed of up to four components: abrasive grains, bond, pores and (depending on the grinding wheel design) wheel hub. The bond links and embeds the abrasive grains. The pores are either naturally created due to the irregular shape and the intergranular space between the abrasive grains or artificially induced by pore builders. The task of the pores is the transport of chips and cutting fluid (Davis 1995, p. 112). The grinding wheel can be classified according to bond type, abrasive material and structure.

The classification of the grinding wheel according to the bond system includes the options of vitreous, organic/synthetic resin, metallically sintered and electroplated bond systems (Davis 1995, p. 110ff.; Klocke 2009, p. 37ff.). With conventional abrasives and super abrasives, two abrasive grain material groups can be distinguished. Conventional abrasives include grains made from black or green silicon carbide and zirconia alumina as well as fused or non-fused aluminium oxide. Super abrasive grains are made either from synthetic diamond or cubic boron nitride (cBN). The hardest abrasive is diamond, followed by cBN, silicon carbide, aluminium oxide and zirconia alumina (Davis 1995, p. 101ff.; Klocke 2009, p. 17ff.). The worldwide production amount of the four abrasive types indicates a broad use of conventional abrasives. The annual production quantity and the mean production energy demand for aluminium oxide, silicon carbide, cubic boron nitride and diamond are presented in Fig. 2.4. In 2013, about 1.19 million tons of aluminium oxide and about 1 million tons of silicon carbide abrasive materials were produced (USGS 2014, p. 13). The amount of super abrasive materials produced was comparably lower, with about 25.1 tons of cubic boron nitride (year 2008) (McClarence 2010, p. 2) and about 33.63 tons of synthetic diamond (year 2005) (Bain and Company 2011, p. 81). The processing of conventional abrasives made of aluminium oxide (purity 99.5 %) demands about 49.5–54.7 MJ/kg and of abrasives made of silicon carbide 70.2–77.6 MJ/kg (Granta Design 2010). The production of super abrasive grains made of cubic boron nitride is connected with an energy demand of 490 MJ/kg (Winter et al. 2015, p. 13). The production of

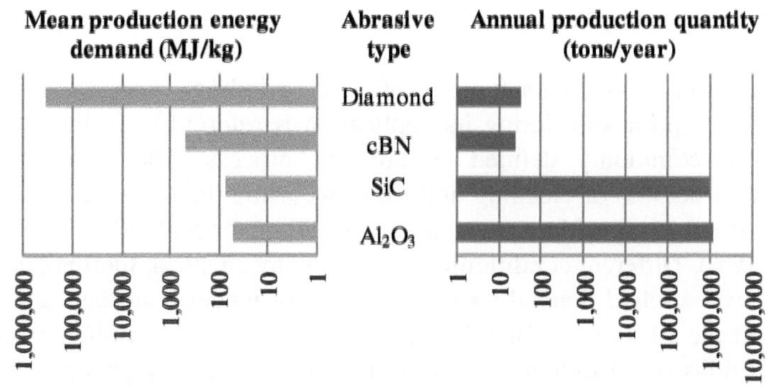

Fig. 2.4 Overview of the mean production energy demand and annual production quantity of different abrasive types (logarithmic scale) (based on USGS 2014, p. 13; McClarence 2010, p. 2; Bain and Company 2011, p. 81; Granta Design 2010; Winter et al. 2015, p. 13; Ali 2011, p. 3)

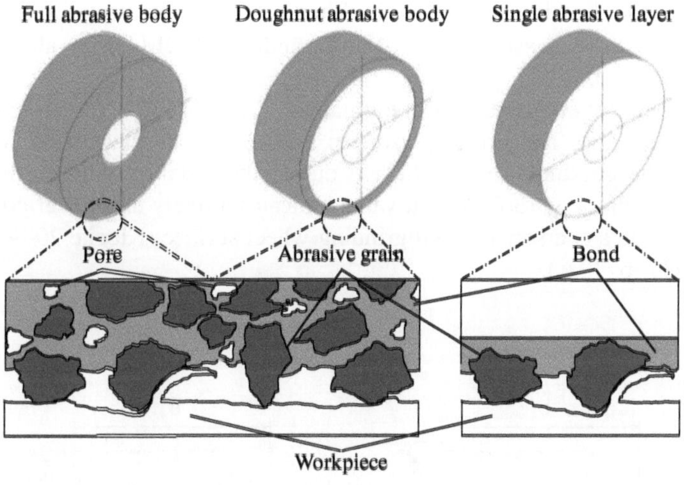

Fig. 2.5 Structure of grinding wheels (Wegener et al. 2011, p. 758)

synthetic diamonds was reported with circa 360 GJ/kg of processing energy (Ali 2011, p. 3).

The grinding wheel structure is divided into a full abrasive body, a doughnut abrasive body and a single abrasive layer (Fig. 2.5). The full abrasive bodies are made of conventional abrasive materials, while super abrasive materials are commonly used in the doughnut-shaped abrasive body or as a single abrasive layer coated on a wheel hub. These variations are due to the availability and costs of the various abrasive materials.

2.1.2.2 Conditioning Process

The preparation and regeneration of the grinding wheel's macro- and micro-geometry prior and during its application is referred to as the conditioning process, also commonly defined as dressing process. The term conditioning includes the processes of cleaning and dressing, while the dressing process can be further classified into profiling and sharpening (Wegener et al. 2011, p. 758; Klocke 2009, p. 135ff.). Different conditioning processes (mechanical, thermal, chemical or hybrid) can be applied depending on the grinding wheel bond and abrasive type (Wegener et al. 2011, p. 757ff:). Usually, mechanical conditioning processes are applied in industrial grinding processes by using a form or profile conditioning kinematic with either fixed or rotating tools (Wegener et al. 2011, p. 759; Saljé 1991, p. 14) (Fig. 2.6).

The mechanical conditioning process presented in Fig. 2.6 is described by the dressing depth (a_{ed}), the dressing tool radius (r_{pd}) and the traverse dressing lead (f_{ad}), as well as the effective dresser width (b_d), the axial dressing overlap (a_{pd}) and the traverse dressing speed (v_{fad}). When a rotating dressing tool is used, the rotational speeds of dressing tool (v_d) and grinding wheel (v_c) are also taken into account (Fig. 2.6).

The dressing overlap ratio can be obtained (U_d) based on the aforementioned variables (Eq. 2.5). The dressing overlap ratio describes the smoothness of the surfaces of the grinding wheel, which is due to the distance between the grooves created by the dressing tool. A high value indicates a finely dressed grinding wheel and a low value a coarsely dressed grinding wheel surface (Klocke 2009, p. 135ff.; Messer 1983, p. 17ff.).

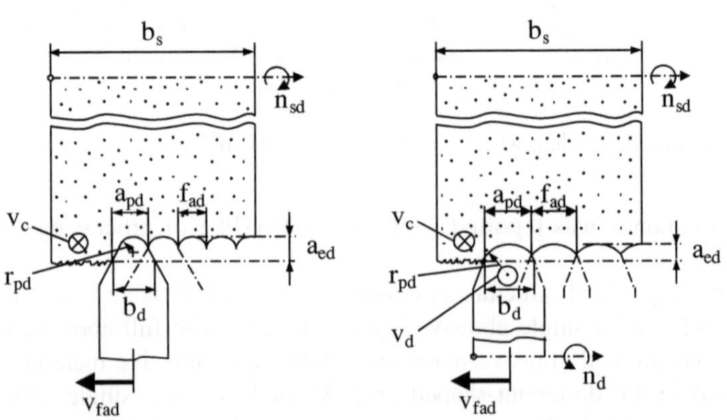

Fig. 2.6 Geometric and kinematic relationships in conditioning processes. **a** Fixed dressing tool. **b** Rotating dressing tool

$$U_d = \frac{b_d}{f_{ad}} \tag{2.5}$$

The dressing speed ratio (q_d) can be calculated by relating dressing tool and grinding wheel speed (Eq. 2.6). The dressing speed ratio influences the length of the engagement path of the dresser in the grinding wheel and therewith the created grinding wheel topography (Klocke 2009, p. 158; Schmitt 1968, p. 55).

$$q_d = \frac{v_d}{v_c} \tag{2.6}$$

2.1.2.3 Cutting Fluid

During the grinding and conditioning process, the application of a cutting fluid is indispensable. The main task of the cutting fluid is to lubricate and/or to cool the processing zone (DIN 51385 2013, p. 4; VDI 3397-2, p. 3f.). A cutting fluid contains two main components: a base fluid and additives. According to DIN 51385, the base fluid is generally divided into non-water miscible and water miscible fluids (Fig. 2.7) (DIN 51385 2013, p. 4).

Non-water miscible cutting fluids are commonly mineral oil-based fluids, enfolding solvent neutrals (paraffinic or naphthenic) and hydrocrack oils. However, vegetable oils and animal fats, containing chemically untreated and treated (ester) fluids, as well as synthetic base oils (polyalphaolefines or polyalkyleneglycos) are also used (Möller and Nassar 2002, p. 72ff.; Silliman 1992, p. 36f.). Water miscible fluids are classified into emulsions and dilutions. An emulsion is a milky and stable disperse mixture, created by the emulsification of mineral base oils, vegetable oil esters or animal fat esters into water. Dilutions generally represent chemically stable and clear fluids composed of water and inorganic and/or organic substances (Möller and Nassar 2002, p. 568ff.; Brinksmeier et al. 1999, p. 583).

Additives are used to improve or constrain specific fluid characteristics (Möller and Nassar 2002, p. 563ff.; Schulze and Holweger 2010, p. 76ff.). Additive-free base fluids can only be used as cutting fluids considering specific restrictions, because the geometry of chip formation as well as high pressures and temperatures prevents a separation of the friction surfaces (Möller and Nassar 2002, p. 563ff.). Additives are necessary to excite physical and chemical reactions, which enable easily shearable boundary layers (Möller and Nassar 2002, p. 563ff.). Accordingly, additives can be classified into enhancers of physical and/or chemical fluid characteristics as well as in tribologically active additives (Hipler et al. 2000, p. 5). Wear protection additives include polar-additives, extreme-pressure-additives and anti-wear-additives. Other additives are agents against corrosion, foaming, oxidation, etc. The additive content in a cutting fluid depends on the application case and varies between a few parts per million up to 20 % of the total (Hipler et al. 2000, p. 6).

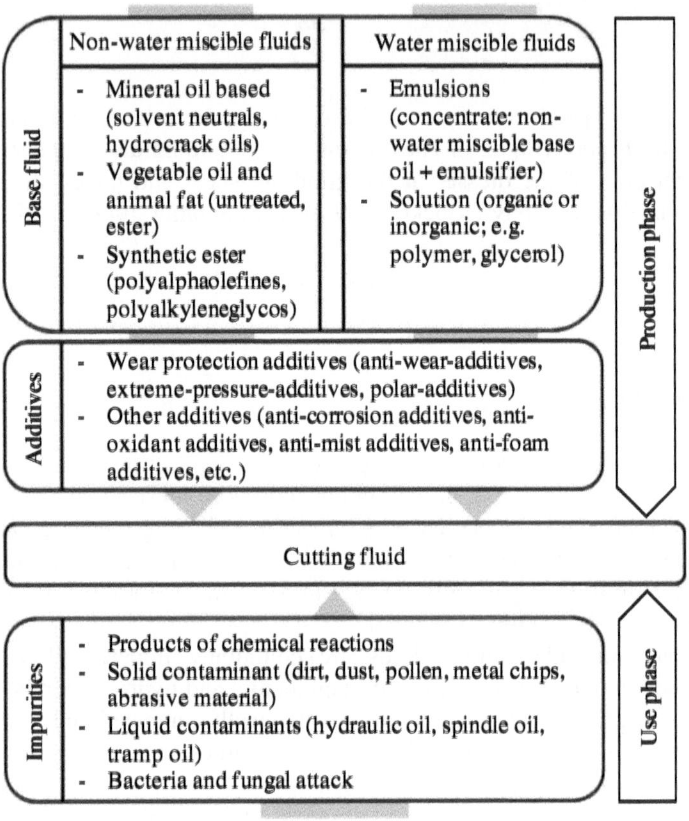

Fig. 2.7 Composition of cutting fluids [according to (DIN 51385 2013, p. 4; Brinksmeier et al. 1999, p. 583)]

The aforementioned base fluids and additives are formulated and mixed for the cutting fluid production according to the required fluid characteristics. However, these characteristics are altered during the utilisation of the cutting fluid, as a result of the demand of additives by the process and the contamination by impurities. These impurities are products of chemical reactions. Examples are the reactions of additives with the workpiece surface during the machining process, detergents adhering to the workpiece or tool surface as well as bacterial or fungal attacks. Further impurities are solid contaminants: this includes produced metal chips and abrasive materials, particles from the environment (dirt, dust, pollen, etc.) as well as organic substances (conglomerates of bacterial colonies and fungal infestations) and liquid contaminants from previous processes (cutting fluid, liquid detergents) or from the machine system (hydraulic oil, slideway oil, etc.) (VDI 3397-2 2007, p. 1ff.; Mang and Dresel 2007, p. 476). A consequence of contamination by these impurities is a change in the original characteristics of the cutting fluid and a reduction of the operating lifetime.

The selection of the cutting fluid depends on tool material, workpiece material, the machining process and process parameters (Vits 1985, p. 95; VDI 3397-1 2007, p. 8f.). According to Vits, the application of non-water miscible cutting fluids in rough grinding processes leads to higher productivity and better surface finishes in comparison with water miscible fluids. In the case of finish grinding, the application of water miscible cutting fluids leads to better results (Vits 1985, p. 95).

In 2013, about 44,800 tons of mineral oil were used to replace or to refill existing cutting fluids in machining processes in Germany. Around 61 % of the total amount were used for non-water miscible cutting fluids. The remaining 39 % were utilised to produce mineral oil-based emulsions (BAFA 2014). Assuming a mineral oil concentration of 5 % (Möller and Nassar 2002, p. 569) led to the production of around 351,500 tons of mineral oil-based emulsion. With an average additive concentration of circa 10 %, around 4,480 tons of additives were needed during the production of cutting fluids.

Procurement prices for cutting fluids range between 3 and 8 €/kg, depending on their formulation with additives, their type and order volume. Considering an average price of 5 €/kg, the sales volume of non-water miscible cutting fluids in Germany is around 136.2 million Euros while that of water miscible cutting fluids is around 87.8 million Euros (year 2013). The total sales volume for mineral oil-based non-water miscible and water miscible cooling lubricants was roughly 805 million Euros in 2008 in the EU27 (Schischke et al. 2012a, p. 6), which amounts to about 161,000 tons of mineral oil-based cutting fluids. These numbers indicate the high economic relevance of the mineral oil-based cutting fluid market.

2.1.3 Grinding System

The grinding system is composed of the grinding machine and its supporting systems. The supporting systems include either a cutting fluid filter or an exhaust air filter (Fig. 2.8). Each of these systems is composed of further subsystems and components, which are described in the next sections.

2.1.3.1 Grinding Machine

Machine tools can generally be defined as energy input dependent stationary systems consisting of linked components, whose professional application results in a product of defined and reproducible geometry (Schischke 2012b, p. 29). The components of a machine tool can be classified into the following items: a machine frame, guides, drives and control units (Tönshoff 1995, p. 16). According to DIN 69651, a grinding machine is a type of machine tool for processes with geometrically undefined cutting edges within the classification of systems for the separation of metal (DIN 69651 1981, p. 3). The design of the grinding machine depends on the grinding process and on the geometry of the machined workpiece,

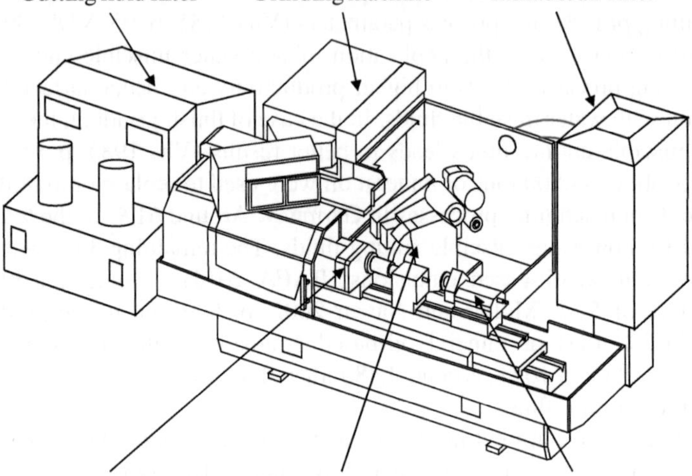

Fig. 2.8 Composition of the grinding system

respectively. Machine types can therefore be classified according to the contact area between workpiece and grinding wheel (peripheral grinding and lateral grinding) or the relative direction of the cutting motion to the main feed motion (down, up, pendular grinding) (Weck and Brecher 2005, p. 219f.).

Around 750.000 CNC machine tools for metal removal processes are installed in the EU27. The distribution of the main types of machine tools (milling machines, turning machines, grinding machines and other cutting machines) is presented in Fig. 2.9. In 2009, the grinding machines had a share of approximately 28 % and represent the second largest group after machining centres (40 %) and before turning machines (23 %) (Schischke et al. 2012c, p. 12ff.). A forecast from Schischke et al. indicates that until 2025 the share of grinding machines will decrease and rank in third place after machining centres and turning machines (Schischke et al. 2012c, p. 12ff.).

2.1.3.2 Cutting Fluid Filter

The cutting fluid filter is an important component in maintaining cutting fluid quality by separating the aforementioned (Fig. 2.7) solid and liquid contaminants from the fluid (VDI 3397-2 2007, p. 2ff.). The process for the separation of these solids can be achieved with five methods: sedimentation, centrifugation, filtration, flotation and magnetic separation (VDI 3397-2 2007, p. 19), while the separation of liquid contaminants can be achieved through skimming, suction, coalescence, flotation and centrifugation (VDI 3397-2 2007, p. 33). During the grinding process,

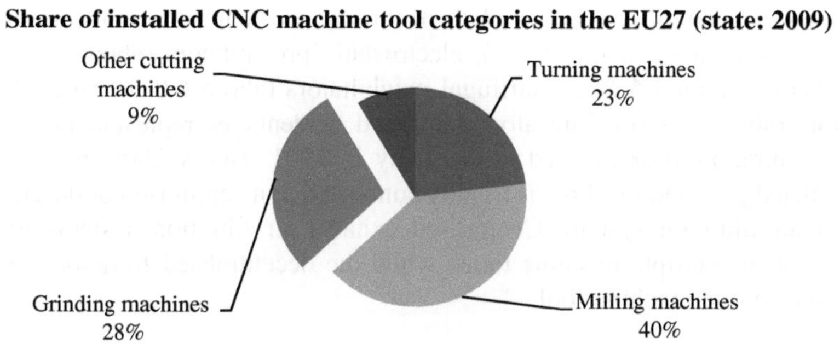

Fig. 2.9 Share of installed CNC machine tools in the EU27 [based on (Schischke et al. 2012c, p. 12ff.)]

the clean cutting fluid is pumped under pressure from a storage tank via the cutting fluid nozzle into the contact zone. After the application the fluid flows back into the cutting fluid filter, is filtered and, if necessary, cooled.

The filter design differs in regard to tank size, number and size of pumps, design of the separation process and the necessity and size of a cooling unit. Furthermore, the grinding process, the number of connected grinding machines and the physical characteristics of the cutting fluid as well as the required fluid purity all have an influence on the cutting fluid filter design (Sutherland 2008, p. 278; VDI 3397-2 2007, p. 18ff.; Silliman 1992, p. 131f.).

Different types of filtration materials, featuring a wet-laid (paper), non-woven and woven structure, can be applied (Sutherland 2008, p. 209ff.). The choice of filtration material depends on the chip size, the cutting fluid type, the cutting fluid filtration principle and the workpiece material. Paper filtration media, for instance, can be applied in a filter cartridge to filter non-water miscible cutting fluids; non-woven filtration material can be used in flat band filtration of water and non-water miscible fluids.

2.1.3.3 Exhaust Air Filtration System

Throughout the grinding process, cutting fluid mist droplets and vapours occur. The mist droplets are between 0.1 and 50 µm (DGUV 2012, p. 20) in size and are mechanically created due to turbulences when the fluid is charged with pressure on the fast rotating grinding wheel and workpiece surface. The cutting fluid vapours are created thermally through the evaporation and incineration of the fluid at high process temperatures. Without the grinding machine encasement and the application of an exhaust air filtration system, the cutting fluid mist droplets and vapours would contaminate the area around the grinding machine and pose a risk to the health of workers and the integrity of the equipment (DGUV 2012, p. 17ff.; BGIA 2006, p. 9ff.).

Exhaust air filters can be divided, according to their filtration concept, into filtering precipitators (about 37 %), electrostatic precipitators (about 50 %), wet precipitators (about 1.5 %), centrifugal precipitators (about 6.0 %) and other precipitators (about 5.5 %). The aforementioned percentages represent the share of different filtration systems used in Germany in 1991 (BGIA 2006, p. 10). Every encapsulated grinding machine is usually connected to a centralised or decentralised exhaust air filtration system. Centralised exhaust air filtration systems filter the exhaust air of multiple machine tools, while the decentralised filtration system is connected to one machine tool.

2.1.4 Energy, Resource and Information Flows of the Grinding Process and System

The grinding process in a grinding system is connected with the demand and transformation of energy, resource and information flows (Fig. 2.10). The input energy flows include the demand and transformation of electrical energy (e.g. powering of the grinding system, realisation of the grinding process), of compressed air (e.g. sealing air inside the tool spindles, workpiece clamping) and of thermal energy (e.g. to cool the spindle or electrical cabinet). Due to the transformation process of energy flows, sonic and vibration emissions as well as thermal emissions from the grinding process and system are created and emitted into the grinding system's environment.

The input resource flows include the unfinished workpiece, the grinding wheel, the conditioning unit and the cutting fluid. The unfinished workpiece is transformed through the application of the aforementioned energy flows into the finished workpiece. In addition to the workpiece, the output resource flows include the chips and tool wear products as well as the grinding wheel and the cutting fluid. The latter ones are only lost for minor proportion and they can be reused as an input resource flow until the end of their service life is reached (Eckebrecht 2000, p. 20f.).

Figure 2.10 shows, besides the energy and resource flows, a third one with the information flows. This flow influences and characterises the relationship between the grinding process and the system. The input information flow includes parameters and state conditions. Among the parameters are the aforementioned geometric and kinematic relationships between the grinding and the conditioning processes, as well as those pertaining to the supply and extraction of the cutting fluid and exhaust air. The state conditions define the characteristics of workpiece, cutting fluid, grinding wheel, conditioning unit, grinding system and supporting units.

The output information flow on the right side encompasses economic, environmental and technological indicators. The economic indicator characterises the cost of the grinding process and system (Eckebrecht 2000, p. 73 ff.; Dettmer 2006, p. 163ff.; Winter et al. 2014, p. 645ff.). The environmental indicators are related to the environmental footprint of the grinding process and system (Dettmer 2006, p. 99ff.; Winter et al. 2014, p. 645ff.). The technological indicators specify the state

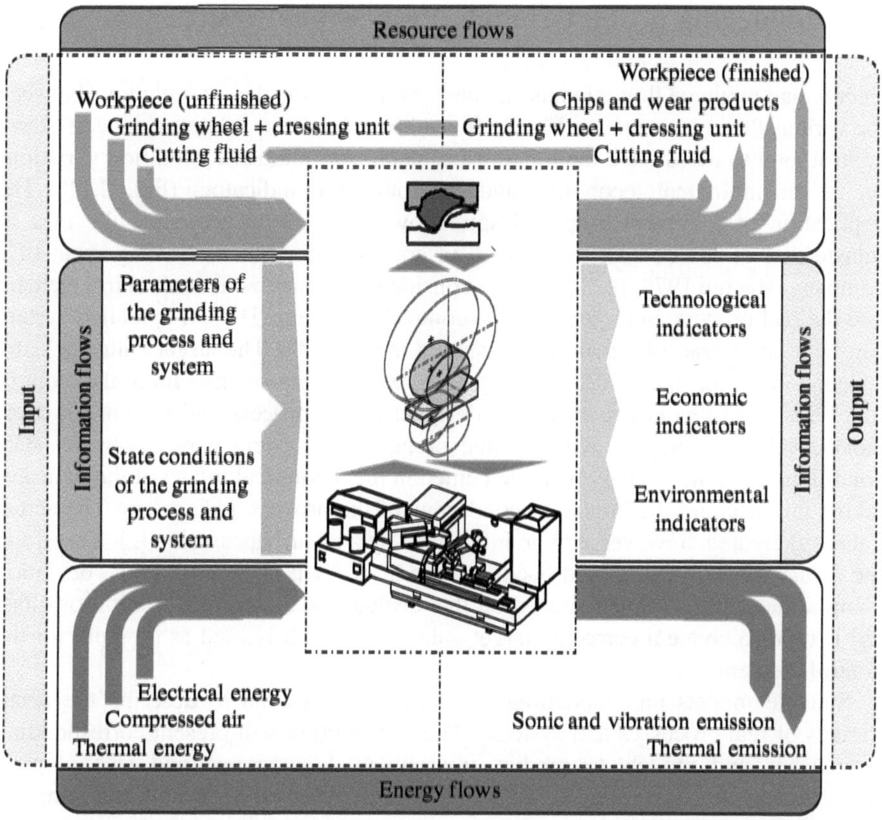

Fig. 2.10 Energy, information and resource flows of the grinding system [based on (Brinksmeier and Schneider 1993, p. 19.06; Winter et al. 2014, p. 645)]

of the ground workpiece (accuracy, surface roughness, material properties, etc.) and the cutting fluid as well as the conditions of the grinding wheel (Rowe 2009, p. 2f.; Winter et al. 2014, p. 645ff.).

The input information flow influences the grinding process and system in the centre. The chosen input information and its achievement potential influences and determines process and system, while the process and system in turn influences the output information flow on the right side in Fig. 2.10. Therefore, changes in the technological, economic and environmental indicators can be achieved by modifying the input information. The range of input information, however, is constrained by limits in the technological, economic and environmental indicators (Winter et al. 2014, p. 645). Furthermore, the indicators partially contradict each other. An increase in cutting depth, for instance, leads to an overall reduction in carbon dioxide emission and costs but also results in an inferior surface quality (Winter et al. 2014, p. 652).

2.2 Modelling of the Grinding Process and System

Energy and resource flow models can be used to describe the interrelations between the various flows presented in Fig. 2.10. Additionally, these models can be used in a control loop to investigate the influence of parameter and state condition variations on the technological, economic and environmental indicators (Fig. 2.11). The application of a control loop is inspired by Baetge, who presented the idea of transferring of the control loop found in control theory to the management of a company (Baetge 1974, p. 30). Baetge considered management as a control element and the real process and system as the controlled system. Thus, a modelled system can be used instead of a real one, as shown in Fig. 2.11. The target value (w) of a specific indicator is defined by the executive management within the control loop. Based on this target value, the designer of the process and system (or operator) develops a strategy with defined process and system parameters and state conditions (u). The strategy is implemented in the theoretical model of the indicator taking into account the boundary conditions of the process and system. The actual value (y) created, however, can be influenced by a disturbance value (z). Therefore, the actual value is transferred back in a feedback loop to determine the deviation using a measuring element. Based on this evaluation, a control deviation or offset (q) is used to create a corrected target value $((f)$ which is used as an input for the control element.

Suitable models and modelling approaches are needed to describe the peculiarities of real processes and systems. The next section will present corresponding approaches and models to predict environmental, economic and technological indicators while considering influencing variables as well as spatial and temporal constraints.

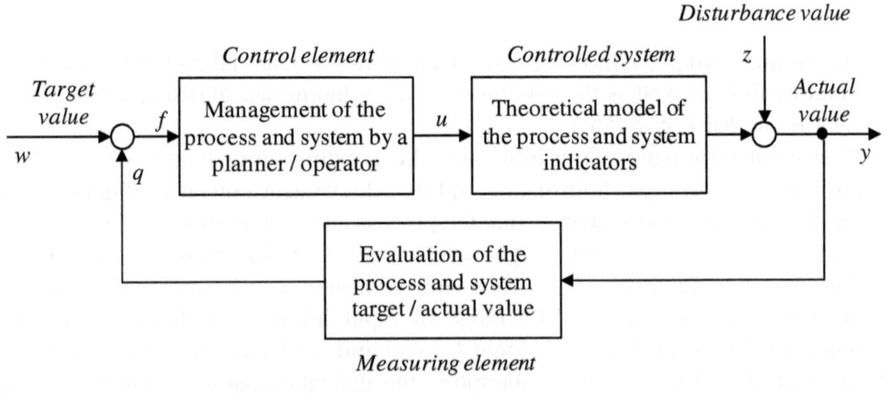

Fig. 2.11 Application of control loop idea to investigate variations of a modelled process and system [based on (Baetge 1974, p. 30)]

2.2.1 Models and Modelling Approaches

The term model can be described as the abstract representation of a process or system, which serves to link causes and effects. It is a mean to reduce the complexity of a process or system to foster understanding and solving (Tönshoff et al. 1992, p. 677; Velten 2009, p. 3). The term possesses three main characteristics: (1) a model is a reproduction of a natural or artificial original; (2) models do not represent all attributes of the original but only those attributes that are relevant to the model creator and/or user; and (3) models are not explicitly associated with their originals (Stachowiak 1973, p. 131ff.).

Different types of models can be distinguished. In case of a model describing a technical system and its relationships, mathematical models can be used. According to Isermann, mathematical models can be derived using a physical or an empirical process analysis (Fig. 2.12) (Isermann 1974, p. 1174). In Fig. 2.12, both types of process analyses are presented. The approach used to derive physical models is presented on the left side, while on the right side is the approach used to derive empirical models.

The physical modelling approach starts with reducing the complexity of the investigated system through the deduction of assumptions for simplification. In the next step, basic equations are derived based on the description of balance equations, physical–chemical relationships, phenomenological equations or entropy balance equations. This set of basic equations of ordinary, partial differential and algebraic equations needs to be solved to derive a first physical model. If the resulting physical model is too complex, it needs to be simplified. A viable option in this context is the reduction of parameters, achieved through the linearisation of equations or by the reduction of the model order (Isermann 1974, p. 1174). Examples of physical models are fundamental analytical models, geometrical kinematic models or models based on molecular dynamics.

The approach used to derive an empirical model is presented in Fig. 2.12 (right side). Experiments are conducted taking into account a priori knowledge about the process (based on a known or unknown structure). During the experiments, the input values as well as the output values are measured. In an identification step, the system relationships are represented as parametric or non-parametric models, depending on whether a model structure can be determined or not. These steps result in an empirical mathematical model (Isermann 1974, p. 1147f.). Examples of empirical models are linear or nonlinear regression models, neural networks and fuzzy logic-based models. Figure 2.12 includes a comparison of physical and empirical modelling. This step can only be carried out if the investigated system is physically and empirically modelled.

Based on the aforementioned comparison, the deviation (difference) between models can be calculated and used to improve each model. Both modelling approaches are an iterative process and the selection of either modelling approach depends on the investigated system. The different characteristics of the empirical and physical modelling approaches are presented in Table 2.1.

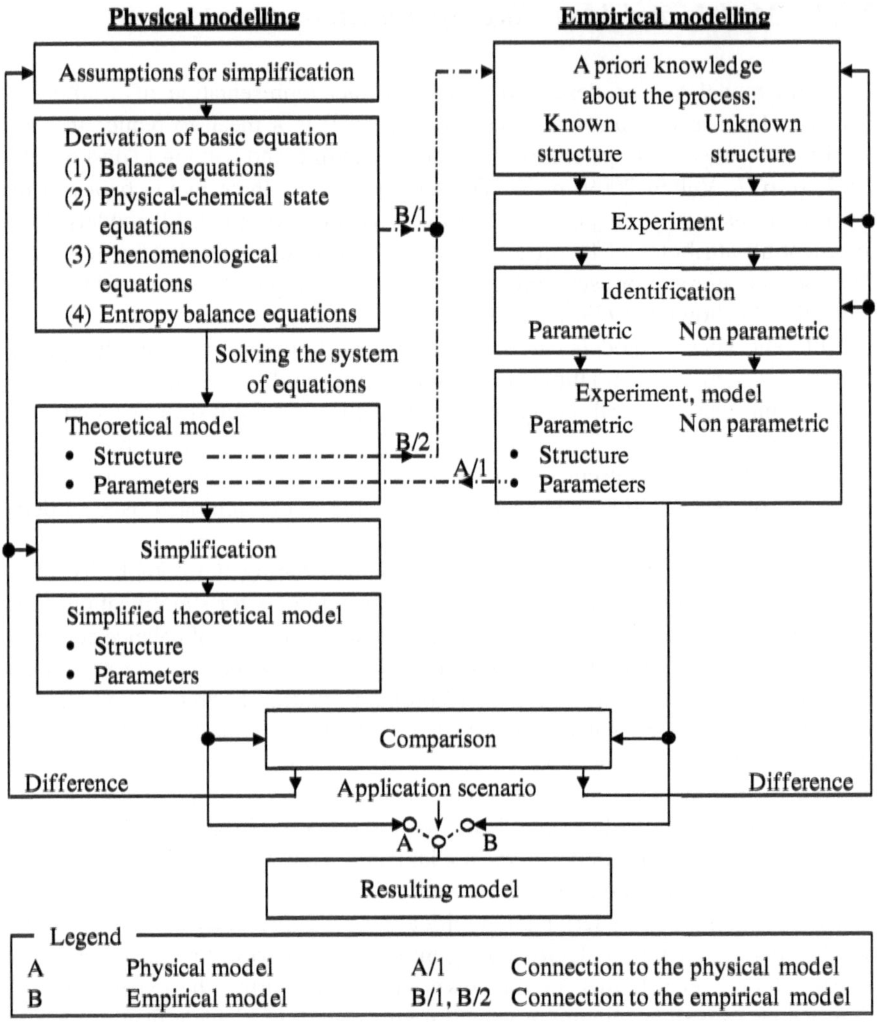

Fig. 2.12 General approach in physical and empirical modelling [according (Isermann 1974, p. 1147)]

The advantage of an empirical approach, in comparison with a physical model, is the precise description of the current system behaviour and a reduced modelling and calculation effort, especially for complex systems. In case of a physical model, all relationships need to be known and mathematically describable. The requirements needed for modelling, calculation and hardware are significantly higher in physical modelling. Furthermore, not all physical relationships are known, and only a limited number of models are available (Knop 1989, p. 18). A disadvantage of the empirical approach, however, is its dependence on experimental data. Therefore, the derived empirical models can only be used for the system being investigated,

Table 2.1 Characteristics of the empirical and physical modelling approaches [based on (Knop 1989, p. 18; Isermann 1974, p. 1148)]

Approach	Physical modelling	Empirical modelling
Characteristics	• Independent from experimental data	• Dependent on experimental data
	• Physical relationships partially unknown, model availability limited	• Limited validity of the model values; yet, very precise description
	• High development and experimental effort before application	• Low development and experimental effort before application
	• High hardware performance necessary	• Low hardware performance necessary
	• If ready for use, then no experiments necessary	• Quick alteration possible, but limited to process
	• Application in science and research	• Application in industrial practice

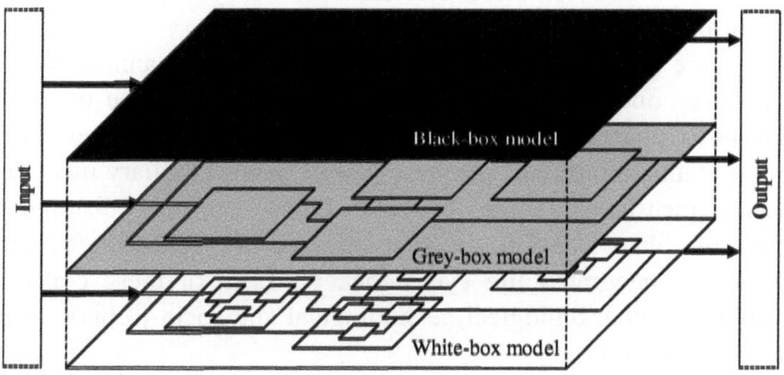

Fig. 2.13 *Black*-box, *grey*-box or *white*-box models

while physical models are case-independent. To overcome these challenges, both modelling approaches can be combined in order to increase the accuracy while reducing the effort (Isermann 1974, p. 1148). The connection is presented in Fig. 2.12 (arrows A/1, B/1 and B/2).

Physical and empirical modelling can be achieved on different levels of aggregation. The level of aggregation can differ between a black-box model, a grey-box model and a white-box model (Fig. 2.13). The black-box model only considers input and output flows entering the investigated system. The system itself, however, is not considered. The white-box-model, on the other hand, takes into account both the flows and all processes and relationships inside the investigated system. The grey-box model is located between the black-box and the white-box models (Haase 2012, p. 27). However, it is difficult to derive a complete white-box model of the investigated system, due to restrictions into the availability of time and data.

Therefore, the derived model can partly have different levels of aggregation (Dettmer 2006, p. 116).

2.2.2 *Spatial and Temporal Dimensions in Modelling*

The modelling of a system comprises a spatial and a temporal dimension. The spatial dimension (introduced in Sect. 2.1.) includes three hierarchical levels of grinding (chip formation, grinding process and grinding system) and their interrelation. These interrelations are the energy and resource flows between the subsystems or levels. Modelling of these relationships can be realised using empirical or physical equations. The temporal dimension can be represented based on the system's life phases to consider its change as well as influence over the time. Accordingly, temporal modelling needs to take into account the energy and resource flows of the different life phases of the included systems, subsystems, components and elements. The spatial dimension is part of one phase of the temporal dimension, but not vice versa. The different life cycle phases are discussed in more detail in the next section.

Life phases are at the core of life cycle concepts (Herrmann 2010, p. 63). The purpose of these concepts is to represent the time dependence of events and to identify particular phases, based on the trend of relevant properties. The analysis, description or evaluation of each of these phases needs auxiliary methods, tools or organisational measures. Three concept types can be distinguished: the flow-oriented life phase concept, the state-oriented life cycle concept and a phase and state-oriented integrated life cycle concept. Each of these concepts can be linked to environmental, biological, technological and socio-technological systems (Herrmann 2010, p. 63).

The flow-oriented life phase concept can be applied to modelling technological systems. This concept illustrates the interrelation of energy and material throughout the life phases of a system (Herrmann 2010, p. 64). The concept can be further divided into a linear and a cyclical life phase as shown in Fig. 2.14.

The linear concept encompasses the following product life phases: raw material extraction, material production, product production, utilisation and disposal. The concept is based on the factual and consequential sequence of the life phases. Every product-related activity can be assigned to a phase (Mateika 2005, p. 11f.). The cyclical concept additionally includes the reutilisation of the product and components as well as physical recycling and natural recycling. As a result, there is a focus on the material and energy flows instead of the product. The concept describes the recycling and reutilisation of materials and energy (Mateika 2005, p. 11f.). The differentiation of physical and natural world allows for the consideration of material and energy flows between both worlds. The figure also shows different temporal modelling dimensions: cradle to grave, cradle to gate, gate to gate and cradle to cradle (Herrmann 2010, p. 64f.).

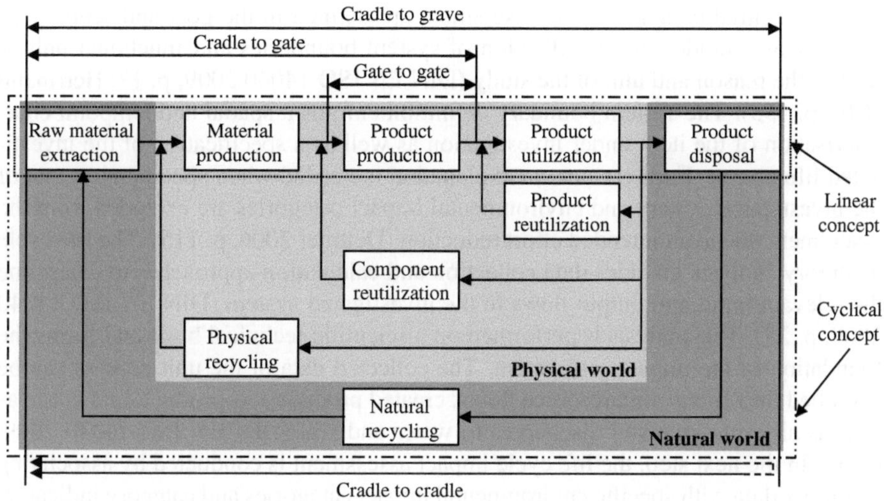

Fig. 2.14 Linear and cyclical product life phase concepts [based on (Herrmann 2010, p. 65)]

Further information on the linear product life phase concept can be found in Senti as well as Stratmann (Senti 1994, p. 10ff.; Stratmann 2001, p. 29ff.) while information on the cyclical product life phase concept can be found in Leber as well as VDI 2243-1 (Leber 1995, p. 45ff.; VDI 2243-1 2002, p. 1ff.).

2.2.3 Modelling of Environmental, Economic and Technological Indicators

The technological, economic and environmental indicators presented in Fig. 2.10 can be used to predict the impact of the grinding process and system on the latter indicators. The prediction of an indicator requires the modelling of such indicator. The following paragraphs will introduce modelling approaches to predict environmental, economic and technological indicators.

The **environmental indicators** can be used to characterise the environmental impact due to energy and resource demand and transformation. The impact can be assessed using the life cycle assessment (LCA) approach found in DIN EN ISO 14040 and 14044. The LCA approach entails the "compilation and evaluation of the inputs, outputs and the potential environmental impacts of a product system throughout its life cycle" (DIN EN ISO 14040 2009, p. 7). The approach comprises four steps: goal and scope definition, inventory analysis, impact assessment and interpretation (DIN EN ISO 14040 2009, p. 17ff.). Due to their iterative nature, it is not necessary to perform the steps in the aforementioned order (Herrmann 2010, p. 153).

The essential contents of the assessment are defined in the goal and scope definition. This includes the specification of system boundaries and functional unit as well as the reason and aim of the study (DIN EN ISO 14040 2009, p. 17; Herrmann 2010, p. 153). The system boundary definition enfolds a spatial and temporal characterisation of the item under investigation as well as a specification of the investigated life phases. Furthermore, an explanation is needed when specific elements of the investigated system and environmental impact categories are excluded from the assessment due to an intended effort reduction (Dettmer 2006, p. 115). The life cycle inventory analysis includes data collection and calculation approaches to determine the relevant input and output flows in the investigated system (DIN EN ISO 14040 2009, p. 25). The analysis is performed on a scientific-technical basis and forms the foundation of the impact assessment. The collected data of the unit process can be classified into energy and resource flows, created products, co-products and waste as well as air emissions and discharges to water and soil (DIN EN ISO 14040 2009, p. 26). In the next step, the life cycle impact assessment is conducted by associating inventory data with specific environmental impact categories and category indicators (DIN EN ISO 14040 2009, p. 27). In addition to selecting impact categories, category indicators and characterisation models, the life cycle impact assessment imperatively includes a classification step as well as a characterisation step (DIN EN ISO 14040 2009, p. 30). In addition to the direct effects, determined via impact categories, indirect effects are also considered, such as health risks due to environmental damages (Herrmann 2010, p. 155). During the interpretation phase, the results of the assessment of the life cycle impact are analysed, evaluated and completed according to previously defined goals and the scope of the investigation (Dettmer 2006, p. 119). For the presentation of the interpreted results, verbal-argumentative, one- or multi-dimensional approaches and methods can be used, such as cumulated energy demand, ECO-Indicator '99, material input per service unit, CML-method (Herrmann 2010, p. 156–163).

The application of the LCA methodology in the modelling and evaluation of the environmental impact of grinding processes and systems was investigated by various authors. The focus on modelling and evaluation varied depending on the investigated element of the grinding process and system. For example, Hansen et al. and Dettmer modelled and evaluated the impact of mineral oil-based cutting fluids in comparison with cutting fluids based upon animal fats and vegetable oil in grinding (Hansen et al. 2005, p. 3ff., Dettmer 2006, p. 99ff.). Furthermore, Dettmer investigated the environmental impact of treatment strategies in the recycling of grinding swarf mixed with different cutting fluid using an LCA (Dettmer 2006, p. 147ff.). In a simplified LCA, Kirsch et al. determined the environmental impact of a vitrified bonded aluminium oxide grinding wheel based on energy demand (Kirsch et al. 2014, p. 74ff.). Linke and Overcash presented a selective overview of the energy and material flows of a grinding process and system and showed options to represent these flows using simplified formulations (Linke and Overcash 2012, p. 293ff.). Murray et al. as well as Winter et al. conducted a simplified LCA of a grinding process and system considering the environmental impact resulting from electrical energy demand, grinding wheel and cutting fluid (Murray et al. 2012,

p. 1604ff.; Winter et al. 2014, p. 644ff.). Dettmer observed a prevalent use of the LCA methodology in research and a minor, but growing, application in industry and politics. The reason for this difference is explained by Dettmer; it is due to the relatively high costs of the assessment in terms of time and money, which affect the realisation of a LCA (Dettmer 2006, p. 120f.).

The created costs and revenues connected with the investigated process and system fall within the **economic indicators**. The costs can be determined through the application of conventional cost accounting (CCA), material flow cost accounting (MFCA) and life cycle costing (LCC).

CCA is part of the internal accounting of companies. Its objective is to provide well-grounded information about the demand of material and about good or service creation, as both are needed to support internal decision-making (Herrmann 2010, p. 136). The CCA can be further divided into cost-type accounting, cost-centre accounting and cost-unit accounting. The approach considers the relationship of time and cost, based on the focus on the past or the future. According to each focus, total absorption costing or incremental costing can be applied (Herrmann 2010, p. 136). Total absorption costing includes the full costs for product or service creation. It rests on the assumption that all costs of a company can be directly assigned. Within incremental costing, only a selected share of the overall cost is directly assigned, while the remaining share is considered as additional overhead cost (Becker et al. 2011, p. 229/251; Herrmann 2010, p. 136). Incremental costing is usually applied in companies, because the underlying organisational structure, including central services (e.g. technical building services), constrains the source-related assignment of costs to a product or service. Costs related to energy, auxiliary and residual materials are allocated as overhead costs and not as specific product or service costs (Schlosser 2013, p. 28). If a critical level is exceeded, the overall operating overhead cost might be beyond control (Schlosser 2013, p. 28).

The MFCA approach, standardised in DIN EN ISO 14051, is a "tool for quantifying the flows and stocks of materials in processes or production lines in both physical and monetary units" (DIN EN ISO 14051 2011, p. 15). Its objective is to enhance the economic and environmental performance of a process thanks to improved material and energy use. The approach allows the identification of production costs and their allocation to the created residual materials. Increasing the transparency of material as well as the energy flows and the transparency of associated costs of products and waste is therefore the main focus of such approach. This transparency allows the calculation of economic saving potential, the identification of hidden costs related to material losses and the prediction of productivity increase, if material losses are prevented (Schmidt 2011, p. 2). Fundamental elements of the MFCA are the quantity centre and the material balance. The quantity centre represents a process or a part of a process. The inputs and outputs of this process are quantified in physical and monetary units. Therefore, the quantity centre serves as a basis for data collection on energy and material flows as well as on material, energy, system and waste management costs (DIN EN ISO 14051 2011, p. 19). The material balance compares the quantity centre's input and outputs to identify missing materials or other gaps in the data. It is based on the assumption that material and energy can neither be destroyed nor created, but only transformed. The application of the MFCA, however, is also connected to restrictions such as the limited consideration of material and recycling loops and the unilateral allocation rule (Viere et al. 2011, p. 652–656).

The LCC approach "…aims at optimising the total costs and yields of a system and of the related activities and processes arising over its life cycle" (VDI 2884 2005, p. 4). Within the aforementioned approach, a supply and demand-oriented perspective can be distinguished (Herrmann 2010, p. 144). The focus of the demand-oriented perspective lies in the determination of costs during the ownership of the investigated element, also called the Total Cost of Ownership. The focus of the supply oriented perspective is on the created revenues or costs in life phases after product sale (Herrmann 2010, p. 144, Dettmer 2006, p. 164). LCC can be achieved either based on approaches suggested by VDI 2884, DIN EN 60300-3-3, VDMA 34160 or in connection with an LCA (Herrmann 2010, p. 145f.; VDI 2884 2005, p. 6ff.; DIN EN 60300-3-3 2014, p. 14; VDMA 34160 2006, p. 2ff.; Dettmer 2006, p. 165). The advantage of the latter option is that the same input and output values can be considered for the calculation of environmental and economic indicators. LCA and LCC are therefore a reliable basis for decision-making (Möller 2000, p. 170).

Various studies have investigated the application of the aforementioned cost models in evaluating the economic impact of grinding processes and systems. For example, Langemeyer as well as Garcia et al., among others, used a simplified CCA to compare different cutting fluid application strategies in surface grinding (flood lubrication with minimal quantity lubrication) (Langemeyer 2002, p. 127ff.; Garcia et al. 2013, p. 106). In combination with a LCA, Dettmer conducted a LCC to compare the impact of a mineral oil-based, a vegetable oil-based and an animal fat-based cutting fluid (Dettmer 2006, p. 162ff.). The LCC methodology was also used by Enparantza et al. to evaluate the economic impact of a grinding process and system (Enparantza et al. 2006, p. 717). A simplified MFCA was conducted by Winter et al. to determine the influence of the cutting fluid type on the economic impact of an internal cylindrical grinding process (Winter et al. 2014, p. 644ff.).

The foundation for the evaluation of environmental and economic indicators is therefore the availability of relevant input and output flows, which can be obtained using an energy and resource flow analysis. Schaltegger and Sturm state that almost every LCA is based on energy and resource flows (Schaltegger and Sturm 2000, p. 65). DIN EN ISO 14051 comes to similar conclusions with regard the costs (DIN EN ISO 14051 2011, p. 19). Thus, the relevant energy and resource flows need to be modelled. For this purpose, the aforementioned physical modelling approach can be used to obtain functions of the input–output relationship. The physical modelling can be realised in an iterative process.

The term **technological indicators** include characteristics, which represent the investigated system or the output of this system with regard to quality as well as energy and resource demand. In case of manufacturing processes, the output is the produced workpiece. The workpiece can be classified according to the achieved quality, such as surface roughness, accuracy and material properties. The demand of

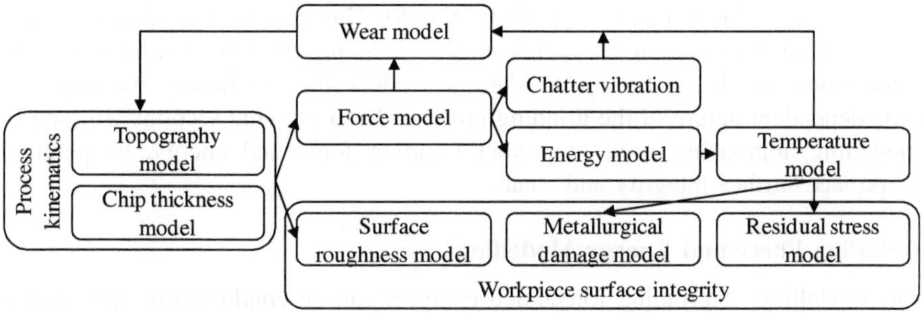

Fig. 2.15 Relationships between physical models in grinding [based on (Brinksmeier et al. 2006, p. 668; Tönshoff et al. 1992, p. 678)]

energy and resources can be classified according to the process and system energy, the wear of the tool or by the forces used during the process. As previously mentioned, the value of each technological indicator depends on the selected input variables. These relationships can be either physically or empirically modelled.

For the physical modelling of the grinding process, the knowledge of the interaction between abrasive grain and workpiece is essential, due to its influence on material impact and the removal process (Brinksmeier et al. 2006, p. 668). Tönshoff et al. (1992, p. 678) and Brinksmeier et al. (2006, p. 668) produced a comprehensive review of the interactions described in the literature and described the relationship between physical models in grinding presented in Fig. 2.15.

The foundation of physical modelling is the knowledge about grinding process kinematics, in the form of grinding wheel topography and chip thickness. The grinding wheel topography model characterises active grain distribution and concentration. Both influence the individual and combined abrasive grain–workpiece interaction and, as a result, the chip formation process or chip thickness. With knowledge about grinding process kinematics, process forces and the achieved workpiece surface roughness can be modelled. A grinding energy model and a wear model can be produced on the basis of the forces. The energy model can be used to derive temperature models for the tribological partner. The temperature model serves as an input for metallurgical damage (e.g. grinding burn) and residual stress modelling. Surface roughness, metallurgical damage and residual stress characterise the surface integrity. The temperature has further impact on the grinding wheel wear model. The wear model itself has an impact on process kinematics, due to the wear of the abrasive grains which change the grinding wheel topography and, as a result, process kinematics.

As defined in Sect. 2.1, the grinding process is characterised by its statistically disorderly located single cutting edge engagement with the workpiece surface. This engagement process needs to be included in process kinematics modelling. Furthermore, the process parameters, such as cutting depth, speeds, feeds, influence the kinematics as well. In addition to the stochastic wheel topography, process kinematic models also need to cover process and time-dependent changes of the grit characteristics (Brinksmeier et al. 2006, p. 668). Thus, physical modelling needs a lot of input information to represent the grinding process. Furthermore, this information needs to be up to date to reflect the stochastic, dynamic and time-dependent nature of the grinding process. The following sections will present modelling approaches used to predict grinding force and energy, temperature, workpiece surface integrity and wear.

Grinding Force and Energy Modelling

The modelling of grinding forces and energies can be conducted on the basis of different approaches (Challen and Oxley 1979, p. 299ff.; Chen 1995, p. 1ff.; Xie and Williams 1996, p. 21ff.; Hou and Komanduri 2003, p. 1579ff.). Challen and Oxley modelled possible interactions of a single hard wedge with a perfectly rigid plastic solid in a two-dimensional approach based on slip-line fields (Challen and Oxley 1979, p. 299 ff.). Using a three-dimensional approach, Xie and Williams describe the interaction between a single and hard pyramid-shaped asperity with a softer surface (Xie and Williams 1996, p. 21ff.). Both approaches are initially proposed in the modelling of wear and friction between two partners in contact and only indirectly focus on force modelling. Badger and Torrance, however, adapted and modified the approaches according to the requirements of the grinding process. Thus, the authors compared both models for the formulation of grinding forces and concluded that the three-dimensional approach leads to more accurate results than the two-dimensional approach (Badger and Torrance 2000, p. 1110f.). In a further investigation, the authors extended the approach and considered the effects of the conditioning process (Torrance and Badger 2000, p. 1787ff.). Hou and Komanduri proposed to model the grinding process mechanics based on probability statistics (Hou and Komanduri 2003, p. 1579ff.). The approach was also used by Rabiey to model a dry grinding process (Rabiey 2010, p. 1 ff.). In this investigation, however, the application of probability and statistics led to limited accuracy (Rabiey 2010, p. 96ff.). A further approach, especially in determining the normal ($f_{n,i}$) and tangential ($f_{t,i}$) force per grain of a grinding process, was suggested by Chen and Rowe. The approach was initially proposed by Shaw and then modified by Chen and Rowe (Chen 1995, p. 1; Chen and Rowe 1996, p. 883 ff.). Connected with the grinding forces are chatter vibrations caused, for example, by unbalanced grinding wheels or the interaction of the grinding wheel with the workpiece (Inasaki et al. 2001, p. 516f.). In order to predict chatter vibrations in grinding processes, Altintas and Weck investigated and presented a comprehensive overview of the physical models (Altintas and Weck 2004, p. 619ff.).

Thermal Modelling

Once sufficient knowledge about the grinding forces and power has been obtained, the workpiece temperature can be modelled. A common reference for most physical temperature models in grinding are the works of Carslaw and Jaeger on a moving heat source (Jaeger 1942, p. 203ff.; Carslaw and Jaeger 2004, p. 1ff). Outwater and

Shaw were the first to apply this model for the thermal analysis of a grinding process. The model was further developed by Hahn, because of the inclusion of the heat source created by the grain wear flat. Takazawa simplified the approach by presenting an approximation equation while DesRuisseaux as well as Zerkle incorporated the impact of the cutting fluid (Brinksmeier et al. 2006, p. 669). A comprehensive model, which takes into account the heat flows into tool, workpiece, cutting fluid and chips, was independently presented by Rowe, Lavine or Malkin and Guo (Rowe 2009, p. 365ff.; Lavine 2000, p. 4447ff.; Malkin and Gou 2008, p. 157ff.).

Workpiece Surface Integrity Modelling

Workpiece surface integrity can be described based on surface roughness, metallurgical damage and residual stress (Comley 2005, p. 24). At the core of physical surface roughness modelling is the assumption that the workpiece surface is created due to the interaction between a hard and sharp asperity and a softer workpiece material (De Vathaire et al. 1981, p. 55ff.; Gilormini and Felder 1983, p. 195ff.). On the basis of this assumption, several authors have described the workpiece surface roughness created by a grinding process based on a similar physical model (Chen and Rowe 1996, p. 883; Torrance and Badger 2000, p. 1787ff.; Nguyen and Butler 2005, p. 1329ff.; Baseri et al. 2008, p. 1091ff.). The approaches differ with regard to the assumed grain shape [sphere shape (Chen and Rowe 1996, p. 883; Nguyen and Butler 2005, p. 1329ff.; Baseri et al. 2008, p. 1091ff.) and triangular shape (Torrance and Badger 2000, p. 1787ff.)]. Rapid cooling and heating during a grinding process can result in residual stress due to phase transformation and thermoplastic deformation of the workpiece material. Connected with the thermally induced residual stress is metallurgical workpiece damage. The term describes changes in micro-hardness and structure, due to thermally induced micro-cracks, fatigue strength or reaustenitisation. The physical modelling of these two processes was investigated by several authors using computer-related algorithms or simulations (Field and Koster 1978, p. 523ff.; Vansevenant 1987, p. 1ff.; Mahdi and Zhang 1998, p. 1289ff.; Moulik et al. 2001, p. 831ff.; Hamdi et al. 2004, p. 277).

Wear Modelling

The superposition of different wear phenomena and mechanisms further complicates physical wear modelling. Meng identified in a literature review of more than 5000 publications (published between 1957 and 1992) about 182 models (Meng 1994, p. 4) to describe the wear process. In this survey, Meng concluded that each investigated model focuses on a different and very small part of the wear process, and no broadly accepted theory exists to fully explain the wear mechanism and phenomena (Meng 1994, p. 79).

Another option, if the physical modelling is complex, inaccurate or time-consuming, is empirical modelling based on a regression analysis of experimental grinding data (Tönshoff et al. 1992, p. 678). Regression analysis describes a functional relationship between the input variables and the output indicators. This relationship usually has an exponential character in grinding and is represented by the general equation shown in (Eq. 2.7) (Knop 1989, p. 25).

$$Y = K_i + C_l \cdot \prod_{i=1}^{n} x_i^{E_i} \tag{2.7}$$

The term Y represents the output indicator and x_i the input variables (usually the process parameters). Each term x_i is quantified by the exponent E_i while the constants K_i and C_j represent the influences that are not explicitly included in the model. An advantage of the exponential model is the representation of a large scope of curved shapes and the determination of the calculated input–output relationship with a small error value, in comparison with the measured values (Knop 1989, p. 26). One of the disadvantages of the equation (Eq. 2.7) is that the description of inflexion points is not possible. This can be of interest if a trend with an opposite behaviour has to be considered, for example time-related influence on the conditioning parameters or grinding parameters (Knop 1989, p. 27). To overcome this drawback, the equation (Eq. 2.7) can be changed to include a superposition of several exponential expressions [see (Eq. 2.8)].

$$Y = K_i + \sum_{l=1}^{m} \left(C_l \cdot \prod_{i=1}^{n} x_i^{E_{li}} \right) \tag{2.8}$$

2.3 Eco-efficiency

2.3.1 Origin of Eco-efficiency

As previously mentioned, the origin of the eco-efficiency approach lies within sustainable development. The sustainable development concept gained an increased importance in the second half of the 20th century. In 1972, Meadows et al. presented the report "The Limits to Growth" which contained a forecast of future development. The report rested on the assumption that the pattern of unsustainable demand and production would have stayed constant for the next decades. The report envisaged a worldwide increased in environmental pollution, depleted natural resources, widespread undernourishment and an increased world population (Meadows et al. 1972, p. 122 ff.). This relationship was also described by Ehrlich and Holdren in (1972). The authors correlated economic development with the created environmental impact and presented the $I = PAT$ equation. The equation relates impact (I) with population (P), affluence (A) and technology (T) (Ehrlich and Holdren 1972, p. 20).

In the following years, this report had a significant influence on the political discussion on sustainability (Herrmann 2010, p. 44). In 1987, a commission under the guidance of Gro Harlem Brundtland presented a first definition of the term sustainable development in the report "Our Common Future". The so-called

Brundtland report defined "sustainable development" as "development that meets the needs of the present without compromising the ability of future generations to meet their own needs" (Brundtland Report 1987, p. 15). In the following years, the concept of sustainable development was widely considered as a goal. As previously mentioned, the concept did not provide any strategies to trigger a change from unsustainable to sustainable development (Mickwitz et al. 2005, p. 1603).

A corresponding approach was first introduced in 1989 by Schaltegger and Sturm through the use of the "eco-efficiency" approach as a means to sustainable development (Schaltegger and Sturm 1989, p. 15ff.). The approach is based on the idea of creating more value for goods and services while creating less waste and pollution. However, it was not until 1992, following the United Nations Conference on Environment and Development held in Rio de Janeiro (also called the "Earth Summit"), that the term became widely published. Prior to the summit, Schmidheiny published the "Changing Course" report as an input to the summit of the Business Council of Sustainable Development (Jollandy 2003, p. 1). The term eco-efficiency was coined with this report (Schmidheiny 1992, 9ff.) and due to its background later described as the "business contribution to sustainable development" (OECD 1998, p. 15). A further achievement of the summit was the "Agenda 21", which proposed suggestions for national policies and strategies to encourage changes in unsustainable demand patterns. Paragraph 4.15 specifies that the achievement of "… the goals of environmental quality and sustainable development will require efficiency in production and changes in consumption patterns to emphasise optimization of resource use and minimization of waste" (UN 1992, p. 20). Therewith, business and politics defined "efficiency" and "sufficiency" as sustainable development strategies in order to separate industrial growth from environmental impact. Both strategies, however, commonly focus only on the improvement of a system, instead of the creation of a new one. To overcome this drawback, a third strategy was introduced: consistency. As previously mentioned, consistency focuses on the reconcilability of economy and environment (Herrmann 2010, p. 51). The consistency strategy is sometimes also labelled with the term eco-effectiveness (Liedtke et al. 2013, p. 1144). According to Schaltegger, the term eco-effectiveness indicates the degree of absolute environmental compatibility of a business activity (Schaltegger 1999, p. 12).

2.3.2 Definition of the Term Eco-efficiency

Eco-efficiency can be defined as a "ratio of some measure of economic value added to some measure of environmental impact. The higher the value added, the more efficient is the use of environmental services" (Ehrenfeld 2005, p. 6). A comprehensive study by Koskela and Vehmas pointed out that there is no commonly agreed or exact definition for the term eco-efficiency (Koskela and Vehmas 2012, p. 547f.). The study further shows that the lack of a common definition is due to different stakeholders using the concept for different purposes, application levels and improvement focuses. As far as the purpose of the application is concerned, the eco-efficiency approach can be used as an assessment method or as a management strategy. Each application level differs according to the political/geographical dimension (regional, national, global), the business dimension (company, corporate, sectoral) and the goods-related dimension (product, service). The improvement focus can be on the productivity increase of the economic value, on the intensity decrease of the environmental impact or on a productivity increase and decrease in intensity at the same time (Koskela and Vehmas 2012, p. 548). From a production perspective, productivity is an absolute measure, defining "the ratio of the output that is produced to the input that it uses" (Coelli et al. 2005, p. 2f.). This definition is sometimes used to define efficiency as well, yet the definitions of productivity and efficiency are not exactly the same. Productivity is an absolute concept based on the ratio of input and output. Efficiency is a relative concept describing the distance between the achieved quantity of input and output, and the optimal quantity of input and output defined by the production frontier (Fried et al. 2008, p. 7f.; Coelli et al. 2005, p. 2f.; Bellgran and Säfsten 2010, p. 260; Daraio and Simar 2007, p. 13f.). In case of eco-efficiency, productivity can be defined as producing more value with less environmental impact. Intensity usually represents the technical performance of a system and is defined as the ratio of produced output per time (Hadeler et al. 2000, p. 1571). In eco-efficiency, intensity can be defined as lowering the environmental impact per created value. Examples of internationally accepted eco-efficiency definitions from governmental and non-governmental organisations are presented in Table 2.2.

Table 2.2 Definition of the term eco-efficiency

World Business Council for Sustainable Development (WBCSD)	"Eco-efficiency is achieved by the delivery of competitively priced goods and services that satisfy human needs and bring quality of life, while progressively reducing ecological impacts and resource intensity throughout the life cycle to a level at least in line with the earth's estimated carrying capacity" (BSCD 1993, p. 9)
Organization for Economic Co-operation and Development (OECD)	"…the efficiency with which ecological resources are used to meet human needs. It can be considered as a ratio of an output divided by an input; the "output" being the value of products and services produced by a firm, a sector or the economy as a whole, and the "input" being the sum of environmental pressure generated by the firm, sector and economy." (OECD 1998, p. 7)
European Environment Agency (EEA)	"A concept and strategy enabling sufficient delinking of the 'use of nature' from economic activity needed to meet human needs (welfare) to allow it to remain within carrying capacities; and to permit equitable access and use of the environment by current and future generations" (EEA 1999, p. 35)
International Standard Organisation (ISO 14045)	"Aspect of sustainability relating the environmental performance of a product system to its product system value" (DIN EN ISO 14045 2012, p. 7)

The first definition in Table 2.2 was formulated in 1993 by the Business Council for Sustainable Development (BCSD) shortly after the Earth Summit. The BCSD merged in 1995 with the World Industry Council for the Environment to form the World Business Council for Sustainable Development (WBSCD). Eco-efficiency can thus be interpreted as a management strategy taking into account the business dimension of a company with a focus on a productivity increase that is inversely proportional to its environmental impact. This definition describes the achievement of the eco-efficiency, when specific economic (competitively priced products) and environmental (earth's estimated carrying capacity) limits are reached (BSCD 1993, p. 9). In 1998, the Organisation for Economic Co-operation and Development (OECD) defined eco-efficiency as a management strategy as well, with a focus on the political/geographical and business dimension. The definition is general and can be interpreted as "more from less"; no limits were specified (OECD 1998, p. 7). The definition formulated by the European Environment Agency (EEA) is a management strategy that can be applied in the political/geographical and business dimensions. The focus is on increasing productivity while lowering the environmental intensity. This is explicitly stated in the passage by describing a "sufficient delinking of the 'use of nature' from economic activity" (EEA 1999, p. 35). Furthermore, a lower limit in regard to the environmental intensity is specified by the carrying capacities of the earth. The last definition in Table 2.2 was published in 2012 by the International Standard Organisation (ISO 14045 and the DIN EN ISO 14045). The definition describes an assessment method to investigate products and services; as a result, there was no specific focus on productivity and intensity (DIN EN ISO 14045 2012, p. 9f.).

2.3.3 *Eco-efficiency Assessment Approaches*

The assessment of eco-efficiency can be conducted using approaches presented by Saling et al. (2002, p. 203ff.), Rüdenauer et al. (2005, p. 105ff.), Steen et al. (2009, p. 1ff.) or the DIN EN ISO 14045 (2012, p. 1f.). The approach presented in DIN EN ISO 14045 represents an internationally accepted standard, which also considers the works of Saling et al. and Rüdenauer et al. The approach presented in DIN EN ISO 14045 will be described in more detail in the following paragraphs. The eco-efficiency assessment approach described in DIN EN ISO 14045 can be applied to investigate the ratio of some economic values in conjunction with the environmental impact of manufacturing processes and systems. The assessment of eco-efficiency comprises five phases: (1) goal and scope definition, (2) environmental assessment, (3) product system value assessment, (4) quantification of eco-efficiency and (5) interpretation.

The **goal and scope definition** (1) specifies the purpose of eco-efficiency assessment, intended audience and intended use of the results, while the scope definition includes a multiple number of items which are needed to specify the key parts of system boundaries, interpretations and limitations (DIN EN ISO 14045 2012, p. 12f.). The description of system boundaries comprises the characterisation of the product system to be assessed in terms of name, scale, location, time and main stakeholders, as well as the function and functional unit of the product system to provide a reference for environmental and value assessment. Furthermore, the system boundary needs to be specified in accordance with ISO 14044 and has to include the definition of the product system value by the description of the value type (functional, monetary or other), as well as the methods used to determine the value. The characterisation of limitations defines the conditions under which the assessment is made, and constrains it to a specific application scenario (DIN EN ISO 14045 2012, p. 12 ff.).

Based on ISO 14040 and ISO 14044, the aforementioned **environmental assessment** (2) is conducted to determine the potential environmental impact of the product system (DIN EN ISO 14045 2012, p. 16 ff.). The calculation is based on the LCA approach and is defined as the "compilation and evaluation of the inputs, outputs and potential environmental impact of a product system throughout its life cycle" (DIN EN ISO 14045 2012, p. 7). The approach comprises the realisation of four steps: goal and scope definition, inventory analysis, impact assessment and interpretation (DIN EN ISO 14040 2009, p. 17 ff.). Due to the iterative nature during the realisation of the assessment approach, it is not necessary to perform the steps in the aforementioned order.

The **product system value assessment** (3) considers the full life cycle of the product system by assessing the worth or desirability of ascribed functional, monetary or other values. The functional value is defined as a "numerical quantity representing functional performance or desirability of a product system, and is subject to improvement" (DIN EN ISO 14045 2012, p. 17). Therefore, the value needs to be measureable. The monetary value of the product system can be reflected in costs, price, added value, etc. Other values can also be considered. These values are characterised by their intangible nature (e.g. aesthetics, brand, cultural and historical values) and can be determined by means of market research (DIN EN ISO 14045 2012, p. 17f.).

During the **quantification of eco-efficiency** (4), the ratio of the results from environmental assessment and product system value assessment is determined in accordance with goal and scope definition. The resulting equation is presented in (Eq. 2.9) (DIN EN ISO 14045 2012, p. 18f.).

$$\text{Eco-efficiency} = \frac{\text{Product system value}}{\text{Environmental impact}} \tag{2.9}$$

In connection with the quantification of eco-efficiency, a sensitivity and uncertainty analysis should be performed. The influence of data and methodology on the calculated eco-efficiency can be determined through a sensitivity analysis. The uncertainty analysis can provide information regarding the influence of data and assumption uncertainties on the result of the assessment.

In the **interpretation** (5) phase, the results of the eco-efficiency assessment are investigated in accordance with the aim of the assessment. Interpretation concludes the eco-efficiency assessment. This step includes the identification of the significant parameters, the evaluation of completeness, sensitivity, uncertainty and consistency and the formulation of conclusions, limitations and recommendations. Foundation of the interpretation phase is the transparent and well-grounded nature of the results (DIN EN ISO 14045 2012, p. 19f.).

2.4 Preliminary Conclusion

As the previous sections have shown, the task of reducing environmental impact while pursuing economic growth and planning or operating an industrial grinding process and system is a complex challenge. Complexity arises from a versatile environment with many competing demands in terms of technological, economic and environmental indicators. Additionally, there are a multitude of combinations for grinding process parameters and system configurations. A definitive statement regarding the influence of the planned or existing grinding process and system on these indicators is not yet available. Accordingly, there is a need for a suitable approach to support decision-making. Such an approach needs to consider the antagonistic interrelationships among the various indicators and the different grinding process parameters and system configurations, in addition to fostering the reduction of environmental and economic impact.

In the light of the existing evidence and specifications, the following conclusions can be drawn:

- In order to represent decision-making scenarios connected with the planning and operating of an industrial grinding process and system, a generic approach is necessary. The approach needs to consider requirements and peculiarities of grinding systems consisting of different subsystems (e.g. grinding machine, supporting unit), components (e.g. grinding wheel, conditioning unit, cutting fluid) and elements (e.g. pump and spindle drives).
- To conduct a straightforward planning of the system and its operation, knowledge about the available and suitable technological, economic and environmental impact indicators as well as their antagonistic relationship is necessary.
- The operation of grinding processes and systems is connected with different energy and resource flows. Each of these flows can be described using process and system parameters as well as properties. To support the planning and operation of grinding processes and systems, it is necessary to specify the relevant flows as well as parameters and properties.
- In addition to the knowledge of the relevant flows, as well as parameters and properties, it is also necessary to determine their influence on the technological, economic and environmental impact indicators. For this purpose, physical and empirical models can be applied. It is necessary that the formulation of the models is accurate and case-independent.
- Manufacturing companies face an increased demand to reduce environmental impact while staying competitive, a demand driven by the political, social and

economic environment of the company. Accordingly, improvement strategies and options are necessary. In addition to strategies and options, it is necessary to provide an evaluation approach, which includes the interrelationships between technological, economic and environmental impact indicators.

The main question based on these concerns is how to cope with the aforementioned challenges. What support to decision-making is necessary when new grinding processes and systems are planned or an existing one needs to be improved?

References

Ali, S.H. (2011): Ecological Comparison of Synthetic versus Mined Diamonds. Working paper, Institute for Environmental Diplomacy and Security, University of Vermont.

Altintas, Y., Weck, M. (2004): Chatter Stability of Metal Cutting and Grinding. CIRP Annals - Manufacturing Technology, 53/2:619–642.

Badger, J.A., Torrance, A.A. (2000): A comparison of two models to predict grinding forces from wheel surface topography. International Journal of Machine Tools and Manufacture, 40/8:1099–1120.

Baetge, J. (1974): Betriebswirtschaftliche Systemtheorie. Regelungstheoretische Planungs-Überwachungsmodelle für Produktion, Lagerung und Absatz. Westdeutscher Verlag, Opladen, Germany.

BAFA (Bundesamt für Wirtschaft und Ausfuhrkontrollen - Referat 423) (2014): Amtliche Mineralöldaten - Entwicklung von Inlandsschmierstoffablieferungen 2006–2013.

Bain & Company (2011): The Global Diamond Industry - Lifting the Veil of Mystery.

Baseri, H., Rezaei, S. M., Rahimi, A., Saadat, M. (2008): Workpiece surface roughness prediction in grinding process for different disc dressing conditions - Part 2: Effects of the wheel topographical parameters on the specific energy and workpiece surface roughness. Machining Science and Technology, 12:197–213.

BCSD (Business Council for Sustainable Development) (1993): Getting Eco-Efficient, Report of the Business Council for Sustainable Development. BCSD, Geneva, Switzerland.

Becker, W., Lutz, S., Back, C. (2011): Gabler Kompaktlexikon Modernes Rechnungswesen. Gabler Verlag, Wiesbaden, Germany.

Bellgran, M., Säfsten, K. (2010): Production Development. Design and Operation of Production Systems. Springer Verlag, London, UK.

BGIA (Berufsgenossenschaftliches Institut für Arbeitsschutz) (Eds.) (2006): Absaugen und Abscheiden von Kühlschmierstoffemissionen. Hauptverband der gewerblichen Berufsgenossenschaften (HVBG), Sankt Augustin, Germany.

Böge, A. (2007): Formeln und Tabellen Maschinenbau. Vieweg Verlag, Wiesbaden, Germany.

Brinksmeier, E., Aurich. J.C., Govekar. E., Heinzel. C., Hoffmeister. H.-W., Peters. J., Rentsch. R., Stephenson. D.J., Uhlmann. E., Weinert. K., Wittmann. M. (2006): Advances in Modeling and Simulation of Grinding Processes. CIRP Annals - Manufacturing Technology, 55/2:311–317.

Brinksmeier, E., Heinzel, C., Wittmann, M. (1999): Friction, Cooling and Lubrication in Grinding. CIRP Annals - Manufacturing Technology, 48/2:581-598.

Brinksmeier, E., Schneider, C. (1993): Bausteine für umweltverträgliche Fertigungsprozesse. Hohe Prozesssicherheit, Hohe Leistung, Hohe Präzision, In: Tagungsband zum 7. Int. Braunschweiger Feinbearbeitungskolloquium, Hrsg.: Westkamper, E., Vulkan Verlag, Essen, Germany.

Brundtland Commission (1987): Our Common Future. Oxford University Press, Oxford, UK.

Carslaw, H., Jaeger, J. C. (2004): Conduction of heat in solids. Oxford Clarendon Press, Oxford, UK.
Challen, J.M., Oxley, P.L.B. (1979): An explanation of the different regimes of friction and wear using asperity deformation models. Wear, 53/2:229–243.
Chen, X. (1995): Strategy for the selection of grinding wheel dressing conditions. Ph.D Thesis, John Moores University, Liverpool, UK.
Chen, X., Rowe, W.B. (1996): Analysis and simulation of the grinding process. Part II: Mechanics of grinding. International Journal of Machine Tools and Manufacture, 36/8:883–896.
Coelli, T., Rao, D.S.P., O'Donnell, C.J. (2005): An Introduction to Efficiency and Productivity Analysis. Springer Science+Business Media, New York, USA.
Comley, P. (2005): Grinding processes and their effects on surface integrity. Ph.D Thesis, Cranfield University, Cranfield, UK.
Daraio, C., Simar, L. (2007): Advanced Robust and Nonparametric Methods in Efficiency Analysis - Methodology and Applications. Springer Science+Business Media, New York, USA.
Davis, J.R. (1995): ASM Specialty Handbook Tool Materials. ASM International, Ohio, USA.
De Vathaire, M., Delamare, F., Felder, E. (1981): An upper bound model of ploughing by a pyramidal indenter. Wear, 66/1:55–64.
Dettmer, T. (2006): Nichtwassermischbare Kühlschmierstoffe auf Basis nachwachsender Rohstoffe. Dr.-Ing. Dissertation, Technische Universität Braunschweig, Vulkan Verlag, Essen, Germany.
DGUV (Deutsche Gesetzliche Unfallversicherung e.V.) (Eds.) (2012): Brand- und Explosionsschutz an Werkzeugmaschinen. BGI/GUV-I 719, Berlin, Germany.
DIN 25424-1 (1990): Fehlerbaumanalyse. Methode und Bildzeichen. Beuth Verlag, Berlin, Germany.
DIN 51385 (2013): Schmierstoffe - Bearbeitungsmedien für die Umformung und Zerspanung von Werkstoffen - Begriffe. Beuth Verlag, Berlin, Germany.
DIN 69651 (1981): Werkzeugmaschinen: Werkzeugmaschinen für die Metallbearbeitung - Begriffe. Beuth Verlag, Berlin, Germany.
DIN 8589-0 (2003): Fertigungsverfahren Spanen. Beuth Verlag, Berlin, Germany.
DIN EN 60300-3-3 (2014): Zuverlässigkeitsmanagement –Teil 3-3: Anwendungsleitfaden – Lebenszykluskosten (IEC 60300-3-3:2004). Beuth Verlag, Berlin, Germany.
DIN EN ISO 14040 (2009): Umweltmanagement - Ökobilanz - Grundsätze und Rahmenbedingungen (ISO 14040:2006). Beuth Verlag, Berlin, Germany.
DIN EN ISO 14045 (2012): Umweltmanagement - Ökoeffizienzbewertung von Produktsystemen - Prinzipien, Anforderungen und Leitlinien (ISO 14045:2012). Beuth Verlag, Berlin, Germany.
DIN EN ISO 14051 (2011): Umweltmanagement - Materialflusskostenrechnung - Allgemeine Rahmenbedingungen. Beuth Verlag, Berlin, Germany.
Eckebrecht, J. (2000): Umweltverträgliche Gestaltung von spanenden Fertigungsprozessen. Dr.-Ing. Dissertation, Universität Bremen, Shaker Verlag, Aachen, Germany.
EEA (European Environment Agency) (1999): Making sustainability accountable: Eco-efficiency, resource producing and innovation. Topic Report No. 11, Copenhagen, Denmark.
Ehrenfeld, J.R. (2005): Eco-efficiency - Philosophy, Theory, and Tools. Journal of Industrial Ecology, 9/4:6–8.
Ehrlich, P.R., Holdren, J.P. (1972): A Bulletin Dialog on "The Closing Circle" Critique. Bulletin of the Atomic Scientists, 28/5:16–27.
Enparantza, R., Revilla, O., Azkarate, A., Zendoia, J. (2006): A Life Cycle Cost Calculation and Management System of Machine Tools. The 13th CIRP International Conference on Life Cycle Engineering, Katholieke Universiteit Leuven, Belgium, May 31 - June 2, 717–722.
Field, M., Koster, W. (1978): Optimizing grinding parameters to combine high productivity with high surface integrity. CIRP Annals - Manufacturing Technology, 27/1:523-526.
Fried, H.O., Lovell, C.A., Schmidt, S.S. (2008): The Measurement of Productive Efficiency and Productivity Growth. Oxford University Press, New York, USA.

Garcia, E., Pombo, I., Sanchez, J.A., Ortega, N., Izquierdo, B., Plaza, S., Marquinez, J.I., Heinzel, C., Mourek, D. (2013): Reduction of oil and gas consumption in grinding technology using high pour-point lubricants. Journal of Cleaner Production, 51:99–108.

Gilormini, P., Felder, E. (1983): Theoretical and experimental study of the plowing of a rigid-plastic semi-infinite body by a rigid pyramidal indenter. Wear, 88/2:195–206.

Granta Design Limited (2010): CES Edupack 2010. Version 6.2.0, Cambridge, UK.

Haase, M. (2012): Entwicklung eines Energie- und Stoffstrommodells zur ökonomischen und ökologischen Bewertung der Herstellung chemischer Grundstoffe aus Lignocellulose. Dr. rer. pol. Dissertation KIT, Karlsruhe, Germany.

Hadeler, T., Winter, E., Arentzen, U. (Ed.) (2000): Gabler Wirtschaftslexikon. Die ganze Welt der Wirtschaft: Betriebswirtschaft, Volkswirtschaft, Recht und Steuern. Gabler Verlag, Wiesbaden, Germany.

Hamdi, H., Zahouani, H., Bergheau, J.-M. (2004): Residual stresses computation in a grinding process. Journal of Materials Processing Technology, 147:277–285.

Hansen, A., Hallmann, C., Schmehl, M. (2005): Produktökobilanz nichtwassermischbarer Kühlschmierstoffe auf Basis von Mineralöl, pflanzlichen Ölen sowie Altspeisefetten und technischen tierischen Fetten. DBU Abschlussbericht AZ 19122, Braunschweig, Germany.

Herrmann, C. (2010): Ganzheitliches Life Cycle Management - Nachhaltigkeit und Lebenszyklusorientierung in Unternehmen. Springer Verlag, Berlin, Germany.

Hipler, F., Gil Girol, S., Fischer, R.A., Wäll, C. (2000): Chemie gegen Reibung und Verschleiß: Untersuchung molekularer Wirkungsmechanismen von Schmierstoffadditiven. Materialwissenschaft und Werkstofftechnik, 31:872–877.

Hou, Z. B., Komanduri, R. (2003): On the mechanics of the grinding process - Part I. Stochastic nature of the grinding process. International Journal of Machine Tools and Manufacture, 43/15:1579-1593.

Inasaki, I., Karpuschewski, B., Lee, H.-S. (2001): Grinding Chatter - Origin and Suppression. CIRP Annals - Manufacturing Technology, 50/2:515–534.

Isermann, R. (1974): Identifikation und Parameterschätzung dynamischer Prozesse. VDI-Z, 116/14:1147–1152.

Jaeger, J.C. (1942): Moving Sources of Heat and the Temperature at Sliding Contacts. Proceedings, Royal Society, New South Wales, 76/3:203–224.

Jollandy, N. A. (2003): An ecological economics of eco-efficiency - theory, interpretations and applications. PhD thesis, Massey University, Palmerston North, New Zealand.

Kirsch, B., Effgen, C., Büchel, M., Aurich, J.C. (2014): Comparison of the embodied energy of a grinding wheel and an end mill. Procedia CIRP, 15:74–79.

Klocke, K. (2009): Manufacturing processes 2 - Grinding, honing, lapping. Springer Verlag, Berlin, Germany.

Knop, M. (1989): Rechnergeführte Stellgrößenwahl beim Außenrundeinstechschleifen - Abrichtereinfluß und Temperaturberechnung. Dr.-Ing. Dissertation, RWTH Aachen, Aachen, Germany.

Koskela, M., Vehmas, J. (2012): Defining Eco-efficiency: A Case Study on the Finnish Forest Industry. Business Strategy and the Environment, 21:546–566.

Langemeyer, A. (2002): Entwicklung und Bewertung von kühlschmierstofffreien Schleifsystemen beim Flachprofilschleifen. Dr.-Ing. Dissertation, Technische Universität Braunschweig, Vulkan Verlag, Essen, Germany.

Lavine, A.S. (2000): An exact solution for surface temperature in down grinding. International Journal of Heat and Mass Transfer, 43:4447–4456.

Leber, M. (1995): Entwicklung einer Methode zur restriktionsgerechten Produktgestaltung auf der Basis von Ressourcenverbräuchen. Dr.-Ing. Dissertation, RWTH Aachen, Shaker Verlag, Aachen, Germany.

Liedtke, C., Buhl, J., Ameli, N. (2013): Microfoundations for Sustainable Growth with Eco-Intelligent Product Service-Arrangements. Sustainability, 5:1141–1160.

Linke, B., Overcash, M. (2012): Life Cycle Analysis of Grinding. Proceedings of the 19th CIRP Conference on Life Cycle Engineering, University of California at Berkeley, Berkeley, USA, May 23-25, 293–298.

Linke, B.S., Dornfeld, D.A. (2012): Application of axiomatic design principles to identify more sustainable strategies for grinding. Journal of Manufacturing Systems, 31:412–419.

Mahdi, M., Zhang, L. (1998): Applied mechanics in grinding - VI. Residual stresses and surface hardening by coupled thermo-plasticity and phase transformation. International Journal of Machine Tools and Manufacture, 38/10-11:1289–1304.

Malkin, S., Guo, C. (2008): Grinding Technology: Theory and Applications of Machining with Abrasives. Industrial Press, New York, USA.

Mang, T., Dresel, W. (2007): Lubricants and Lubrication. WILEY-VCH Verlag, Weinheim, Germany.

Marinescu, I.D., Hitchiner, M., Uhlmann, E., Rowe, W.B., Inasaki, I. (2007): Handbook of machining with grinding wheels. CRC Press, Taylor & Francis Group, Florida, USA.

Mateika, M. (2005): Unterstützung der lebenszyklusorientierten Produktplanung am Beispiel des Maschinen- und Anlagenbaus. Dr.-Ing. Dissertation, Technische Universität Braunschweig, Vulkan Verlag, Essen, Germany.

McClarence, E. (2010): Report - Trends for cubic boron nitride. The Abrasives Hub.

Meadows, D.H., Meadows, D.L., Randers, J., Behrens III, W.W. (1972): The limits to growth. Universe Books, New York, USA.

Meng, H.-C. (1994): Wear Modeling, Evaluation and categorization of wear models. Ph.D Thesis, University of Michigan, Ann Arbor, USA.

Messer, J. (1983): Abrichten konventioneller Schleifscheben mit stehenden Werkzeugen. Dr.-Ing. Dissertation, RWTH Aachen, Aachen, Germany.

Mickwitz, P., Melanen, M., Rosenström, U., Seppälä, J. (2005): Regional eco-efficiency indicators - a participatory approach. Journal of Cleaner Production, 14:1603–1611.

Möller U.J., Nassar J. (2002): Schmierstoffe im Betrieb. Springer Verlag, Berlin, Germany.

Möller, A. (2000): Grundlagen stoffstrombasierter betrieblicher Umweltinformationssysteme. Dr. rer.nat. Dissertation, Univeristät Hamburg, Projekt Verlag, Bochum, Germany.

Moulik, P.N., Yang, H.T.Y., Chandrasekar, S. (2001): Simulation of thermal stresses due to grinding. International Journal of Mechanical Sciences, 43:831–851.

Murray, V.R., Zhao, F., Sutherland, J.W. (2012): Life cycle analysis of grinding: a case study of non-cylindrical computer numerical control grinding via a unit-process life cycle inventory approach. Proceedings of the Institution of Mechanical Engineers, Part B: Journal of Engineering Manufacture, 226/10:1604–1611.

Nguyen, T.A., Butler, D.L. (2005): Simulation of surface grinding process, part 2: interaction of the abrasive grain with the workpiece. International Journal of Machine Tools and Manufacture, 45/11:1329–1336.

OECD (Organization for Economic Co-operation and Development) (1998): Eco-efficiency, Paris, France.

Rabiey, M. (2010): Dry Grinding with cBN Wheels - The effect of structuring. Dr.-Ing. Dissertation, Universität Stuttgart, Jost-Jetter Verlag, Stuttgart, Germany.

Ropohl, G. (2009): Allgemeine Technologie - Eine Systemtheorie der Technik. Universitätsverlag Karlsruhe, Karlsruhe, Germany.

Rowe, W.B. (2009): Principles of modern grinding technology. William Andrew, Elsevier, Oxford, UK.

Rüdenauer, I., Gensch, C.-O., Grießhammer, R., Bunke, D. (2005): Integrated Environmental and Economic Assessment of Products and Processes - A Method for Eco-Efficiency Analysis. Journal of Industrial Ecology, 9/4:105–116.

Saling, P., Kicherer, A., Dittrich-Kraemer, B., Wittlinger, R., Zombik, W., Schmidt, I., Schrott, W., Schmidt, S. (2002): Eco-efficiency Analysis by BASF: The Method. The International Journal of Life Cycle Assessment, 7/4:203–218.

Saljé, E. (1991): Begriffe der Schleif- und Konditioniertechnik. Vulkan Verlag, Essen, Germany.

Saljé, E., Möhlen, H. (1986): Fundamental Dependecies upon Lenghts and Results in Grinding. CIRP Annals - Manufacturing Technology, 35/1:249–253.
Schaltegger, S. (1999): Öko-Effizienz als Element des sozio-ökonomisch vernünftigen Umweltmanagements. Ökologisches Wirtschaften, 3:12–14.
Schaltegger, S., Sturm, A. (1989): Ökologieinduzierte Entscheidungsprobleme des Managements: Ansatzpunkte zur Ausgestaltung von Instrumenten. WWZ-Discussion Paper No. 8914.
Schaltegger, S., Sturm, A. (2000): Ökologieorientierte Entscheidungen in Unternehmen. Paul Haupt, Bern, Switzerland.
Schischke, K., Hohwieler, E., Feitscher, R., König, J., Kreuschner, S., Wilpert, P., Nissen, N.F. (2012a): Energy-Using Product Group Analysis - Lot 5, Machine tools and related machinery, Executive Summary – Final Version, Berlin, Germany.
Schischke, K., Hohwieler, E., Feitscher, R., König, J., Kreuschner, S., Wilpert, P., Nissen, N.F. (2012b): Energy-Using Product Group Analysis - Lot 5, Machine tools and related machinery, Task 1 Report – Definition, Berlin, Germany.
Schischke, K., Hohwieler, E., Feitscher, R., König, J., Kreuschner, S., Wilpert, P., Nissen, N.F. (2012c): Energy-Using Product Group Analysis - Lot 5, Machine tools and related machinery, Task 2 Report – Economic and Market Analysis, Berlin, Germany.
Schlosser, R. (2013): Methodik zur Prognose der Nachhaltigkeit des Energie- und Stoffeinsatzes spanender Fertigungsprozesse. Dr.-Ing. Dissertation, RWTH Aachen, Shaker Verlag, Aachen, Germany.
Schmidheiny, S. (1992): Changing course: A global business perspective on development and the environment. MIT Press, Cambridge, USA.
Schmidt, M. (2011): Materialflusskostenrechnung. RKW Rationalisierungs- und Innovationszentrum der Deutschen Wirtschaft e.V. Kompetenzzentrum, Faktenblatt: 2/2011.
Schmitt, R. (1968): Abrichten von Schleifscheiben mit diamantbestückten Rollen. Dr.-Ing. Dissertation, Technische Universität Braunschweig, Braunschweig, Germany.
Schulz, J., Holweger, W. (2010): Wechselwirkung von Additiven mit Metalloberflächen. Expert Verlag, Renningen, Germany.
Senti, R. (1994): Produktlebenszyklusorientiertes Kosten- und Erlösmanagement. Dr. oec. HSG Dissertation, Universität St. Gallen, Difo-Druck, Bamberg, Germany.
Silliman, J.D. (Ed.) (1992): Cutting and grinding fluids: Selection and application. Society of Manufacturing Engineers, Dearborn, USA.
Stachowiak, H. (1973): Allgemeine Modelltheorie. Springer Verlag, Wien, Austria.
Steen, B., Carlson, R., Lyrstedt, F., Skantze, G. (2009): Sustainability management of businesses through eco-efficiency - an example. CPM Report, 2009:3.
Stratmann, J. (2001): Bedarfsgerechte Informationsversorgung im Rahmen eines produktlebenszyklusorientierten Controlling. Dr.-Ing. Dissertation, Technische Universität Berlin, Josef Eul Verlag, Köln, Germany.
Sutherland, K. (2008): Filters and filtration handbook. Butterworth-Heinemann, Oxford, UK.
Tönshoff, H.K. (1995): Werkzeugmaschinen: Grundlagen. Springer Verlag, Berlin, Germany.
Tönshoff, H.K., Peters, J., Inasaki, I., Paul, T. (1992): Modelling and Simulation of Grinding Processes. CIRP Annals - Manufacturing Technology, 41/2:677–688.
Torrance, A.A., Badger, J.A. (2000): The relation between the traverse dressing of vitrified grinding wheels and their performance. International Journal of Machine Tools and Manufacture, 40/12:1787–1811.
UN (United Nations) (Ed.) (1992): Earth Summit. Agenda 21. The United Nations Programme of Action from Rio. The final text of agreements negotiated by Governments at the United Nations Conference on Environment and Development (UNCED).
USGS (U.S. Geological Survey) (2014): Mineral commodity summaries 2014. U.S. Geological Survey, Reston, USA.
Vansevenant, E. (1987): A Subsurface Integrity Model in Grinding. Dr.-Ing. Dissertation, Katholieke Universiteit Leuven, Leuven, Belgium.
VDI 2243 (2002): Recyclingorientierte Produktentwicklung. VDI-Handbuch Konstruktion, VDI-Handbuch Umwelttechnik, Beuth Verlag, Berlin, Germany.

VDI 2884 (2005): Beschaffung, Betrieb und Instandhaltung von Produktionsmitteln unter Anwendung von Life Cycle Costing (LCC). VDI-Handbuch Betriebstechnik, Teil 4, Beuth Verlag, Berlin, Germany.

VDI 3397-1 (2007): Kühlschmierstoffe für spanende und umformende Fertigungsverfahren - Blatt 1. VDI-Handbuch Betriebstechnik, Teil 4, Beuth Verlag, Berlin, Germany.

VDI 3397-2 (2007): Kühlschmierstoffe für spanende und umformende Fertigungsverfahren - Blatt 2. VDI-Handbuch Betriebstechnik, Teil 4, Beuth Verlag, Berlin, Germany.

VDMA 34160 (2006): Prognosemodell für die Lebenszykluskosten von Maschinen und Anlagen. Beuth Verlag, Berlin, Germany.

Velten, K. (2009): Mathematical modeling and simulation - introduction for scientists and engineers. WILEY-VCH Verlag, Weinheim, Germany.

Viere, T., Prox., M., Möller, M., Schmidt, M. (2011): Implications of material flow cost accounting for life cycle engineering. Proceedings of the 18th CIRP International Conference on Life Cycle Engineering, Technische Universität Braunschweig, Braunschweig, May 2-4, 652–656.

Vits, R. (1985): Technologische Aspekte der Kühlschmierung beim Schleifen. Dr.-Ing. Dissertation, RWTH Aachen, Aachen, Germany.

Weck, M., Brecher, C. (2005): Werkzeugmaschinen - Maschinenarten und Anwendungsbereiche. Springer Verlag, Berlin, Germany.

Wegener, K., Hoffmeister, H.-W., Karpuschewski, B., Kuster, F., Hahmann, W.C., Rabiey, M. (2011): Conditioning and monitoring of grinding wheels. CIRP Annals - Manufacturing Technology, 60/2:757–777.

Werner, G. (1971): Kinematik und Mechanik des Schleifprozesses. Dr.-Ing. Dissertation, RWTH Aachen, Aachen, Germany.

Winter, M., Ibbotson, S, Kara, S., Herrmann, C. (2015): Life cycle assessment of cubic boron nitride grinding wheels. Journal of Cleaner Production, 107:707–721.

Winter, M., Li, W., Kara, S., Herrmann, C. (2014): Determining optimal process parameters to increase the eco-efficiency of grinding processes. Journal of Cleaner Production, 66:644–645.

Xie, Y., Williams, J.A. (1996): The prediction of friction and wear when a soft surface slides against a harder rough surface. Wear, 196/1-2:21–34.

Chapter 3
State of Research

Abstract The state of research is reviewed in this chapter. The review comprises current research approaches in describing, modelling, evaluating and improving the economic, environmental and technological impact of machining processes. These approaches are comprehensively evaluated with the aim of providing a comparative overview. The chapter highlights the need for further research on how to describe, mode, evaluate and to improve eco-efficiency.

3.1 Review of Research Approaches

In order to answer the questions that have been introduced in the previous chapters, this section presents a variety of research approaches used to support decision-making during the planning of a new or the improvement of an existing manufacturing process and system. The approaches differ depending on their focus on the technological, economic and/or environmental assessment as well as their application level. Due to the great variety of existing approaches, this review will focus on research investigations which:

- describe, model, evaluate or improve machining processes with geometrically defined and undefined cutting edge;
- focus not only on technological assessment, but consider the economic and/or the environmental assessment as well; and
- present an original research approach and contribute to the achievement of sustainability, eco-efficiency, resource efficiency, energy efficiency or environmentally benign manufacturing.

The review classifies the investigations according to their hierarchical level and approach focus. The hierarchical level describes on which level the investigation was conducted and the system's response. In accordance with Fig. 2.1, the second and third hierarchical levels and a combination of both are considered. However, due to the review of investigations with geometrically defined and undefined cutting edge, general terms for the second and third level were chosen instead of terms

M. Winter, *Eco-efficiency of Grinding Processes and Systems*,
Sustainable Production, Life Cycle Engineering and Management,
DOI 10.1007/978-3-319-25205-6_3

specific for the grinding process. The hierarchical level is subdivided into process level, system level as well as process and system level. The process level represents, in accordance with Fig. 2.1, the second hierarchical level and considers relationships within the tool–workpiece–cutting fluid interaction. The system level represents the third hierarchical level of Fig. 2.1 and considers the interaction between the machine tool and the support units. The process and system level represents a combination of process level and system level.

The approach focus is subdivided into four categories, depending on whether they describe, model, evaluate and/or improve.

- The first category (I) include approaches that describe the respective system level based on the application of common methods (e.g. LCA, LCC) and procedural approaches (e.g. machining experiments).
- The second category (II) entails approaches that describe and model a system. This includes the development of a new model to describe a system or the use of the description to derive a model of a system.
- The third category (III) comprises approaches, which include description, modelling and evaluation steps to determine the impact of changes in the corresponding hierarchical level.
- The last category (IV) is used to classify approaches, which can be used to in describing, modelling, evaluating and improving a system.

In total, 70 approaches to support decision-making during the planning and improvement of processes and systems were identified and classified into a portfolio as shown in Fig. 3.1. The horizontal axis (x-axis) of the portfolio presents the three hierarchical levels and the vertical axis (y-axis) the four approach categories.

Furthermore, each area in the portfolio indicates with a black bullet point whether the approach focuses on a technological, environmental and/or economic assessment. Approaches, which address grinding processes, are highlighted in bold letters in the figure.

The portfolio indicates a high number of research studies, which focus on the process level, followed by the combination of process and system levels and the system level. As far as the approach categories are concerned, the category (I) approaches are the majority, followed by category (III) and category (IV), while the category (II) approaches are fewest in number. Approaches focusing on the technological assessment dominate, especially on a process level, in the category (I) as well as category (IV). The three areas in the upper right-hand corner of the portfolio show a predominance of investigations based on environmental and technological assessment. For the determination of the environmental and economic indicator, a system-based perspective is necessary. When determining a technological indicator, a process-based perspective is sufficient. Furthermore, the majority of the investigations focus on the impact of machining processes with geometrically defined cutting edges, especially at the system level and the combination of process and system level. Processes with geometrically undefined cutting edge have a higher share on a process level.

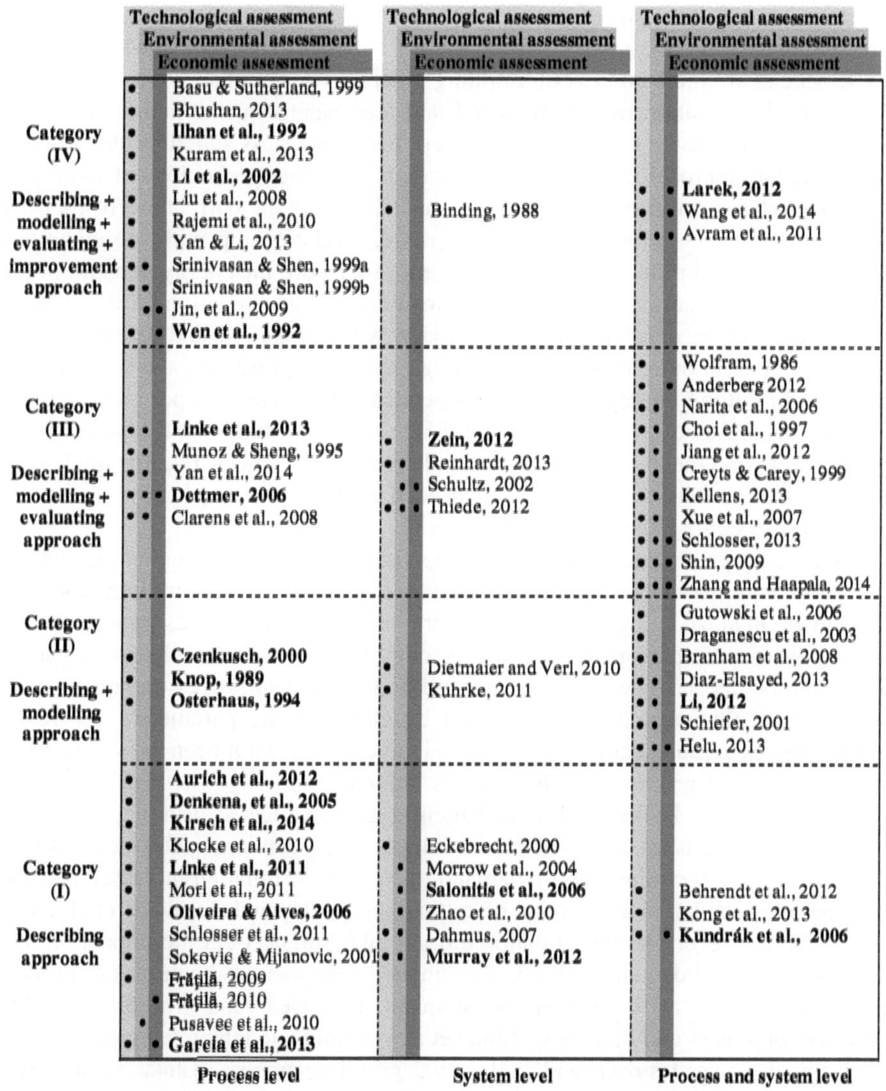

Fig. 3.1 Classification of identified research approaches with regard to investigation level and scope of the applied approach (approaches with focus on grinding in bold)

3.1.1 *Process Level*

Category (I)—Describing Approaches

Approaches describing the process level can generally be classified according to whether the focus of the investigation is on the impact of process parameter variation or tool–workpiece–cutting fluid interaction. The technological impact of process parameter variation on power demand in the cutting process was described by Klocke et al. in the context of a milling process (Klocke et al. 2010, pp. 346ff.). Based on the variation of cutting speed and feed per tooth, the authors aimed at identifying the most energy-efficient process parameters. As stated by the authors, however, the approach lacks the consideration of further technological indicators such as tool wear. Information on tool wear allows for the correlation of tool life with selected process parameters. Furthermore, a relationship with the embodied energy during tool production can be created; as a result, the energy demand of the process and the energy demand of tool manufacturing in relation to the process parameters can be considered. The approach was transferred and extended from Schlosser et al. in order to analyse a drilling process by presenting the behaviour of the specific cutting energy demand in dependence with the feed per tooth and the cutting speed (Schlosser et al. 2011, pp. 85ff.). The investigation still lacked focus on tool wear and the needed energy for tool production. However, a first qualitative description regarding the impact of the needed energy in the production of the tool in relation to cutting speed was presented.

Mori et al. presented a study describing the process energy demand in cutting processes with geometrically defined cutting edge (drilling, face milling, end milling) (Mori et al. 2011, pp. 145ff.). The energy demand of the process was measured according to different process parameter combinations. The results of the study indicated the dependence of power demand on the selected process parameters and that process efficiency can be increased by changing the parameters. However, further values to represent the technological indicator have not been presented; thus, the evaluation of process eco-efficiency is limited.

Denkena et al., Linke et al. and Aurich et al. investigated grinding processes. A study by Denkena et al. showed the influence of process parameter variations on cutting power and cutting energy demand and concluded that an increased specific material removal led to a reduction of cutting energy demand (Denkena et al. 2005, pp. 39ff.). However, the investigation did not take into account the impact on tool wear, cutting fluid demand or workpiece quality. Considering the impact of process parameter change on cutting energy demand, tool wear and workpiece quality (surface roughness and hardness), Linke et al. presented an integrated technological analysis of a speed stroke and high-speed grinding process (Linke et al. 2011, pp. 1). The study did not include any environmental or economic analyses. Aurich et al. described the influence of process parameter variation on the specific process energy demand and the achieved surface roughness (Aurich et al. 2012, pp. 329ff.). The influence of grinding wheel specification and grinding direction was also investigated. The presented results were obtained in a round robin test within "The International Academy for Production Engineering" (CIRP), which led to the investigation of different grinding processes and conditions. The experimental conditions and a further description were presented in a later publication (Aurich et al. 2013, pp. 653ff.). No information on the influence of process parameter variation on environmental and economic indicators was presented.

Studies by Soković and Mijanović, Frăţilă and Pušavec et al. investigated the influence of the cutting fluid in cutting processes with geometrically defined cutting edges. In a technological investigation, Soković and Mijanović analysed the influence of three different cutting fluids in a multipoint tapping process (Soković and Mijanović 2001, pp. 181ff.). During the investigation, only the cutting speed was modified, and the main focus was on the technological influence of parameter variation. The environmental and the economic indicators were not investigated. By looking at the technological and environmental indicators, Frăţilă focused on the influence of a near-dry and a flood machining cutting fluid supply strategy in gear milling. The technological study of the workpiece surface roughness and the hob wear showed that the application of the flood cutting fluid supply led to better surface roughness as well as higher tool wear, compared to the near-dry fluid supply (Frăţilă 2009, pp. 839ff.). An environmental analysis of both cutting fluid supply strategies was conducted as a following step (Frăţilă 2010, pp. 1031ff.). However, the environmental comparison of both strategies was partially limited, due to the application of different tools when using each cutting fluid supply strategy. Furthermore, the life cycle inventory was based on the produced steel scrap quality, the electrical energy demand and the produced and disposed of cutting fluid. Tool impact was not considered. Based on these data, the application of the near-dry cutting fluid strategy resulted in a lower environmental impact than the flood strategy. In an environmental investigation, Pušavec et al. compared a turning process with conventional flood cutting fluid supply, with a high pressure jet-assisted machining (HPJAM) and a cryogenic machining process (Pušavec et al. 2010, pp. 174ff.). The environmental impact analysis was performed using an LCA. The system boundary of the study only took into account the production and supply of the cutting fluids into the machining process. Therefore, the results are partially limited due to the lack of focus on cutting fluid use and disposal phase. The investigation indicated a higher environmental impact of the conventional cutting fluid strategy compared to HPJAM and cryogenic machining.

In a technological study, Oliveira and Alves compared the impact of different cutting fluid strategies on tool wear, normal force and surface roughness in a grinding process (Oliveira and Alves 2006, pp. 343ff.). A castor oil-based cutting fluid was compared with a semi-synthetic fluid. The application of the castor oil-based fluid showed better technological results than the semi-synthetic fluid. However, the investigation lacked information about the environmental and economic impact of such strategies. In an investigation, combining technology and economics, Garcia et al. studied the impact of cutting fluid supply strategies on a cylindrical grinding process (Garcia et al. 2013, pp. 99ff.). The main focus of the investigation was the influence of a conventional flood cutting fluid supply strategy in comparison with a combination of minimal quantity lubrication and cryogenic cooling. Both strategies were compared based on the specific cutting energy, the tool wear and the residual stress within the workpiece. Both from a technological and an economic perspective, the application of the minimal quantity lubrication with cryogenic cooling led to better results than the flood cutting fluid supply strategy. However, the study lacked information on the environmental impact resulting from the production, use and disposal phases of the cutting fluid supply strategies.

Kirsch et al. compared the overall embodied energy of grinding and milling of Inconel 718 (Kirsch et al. 2014, pp. 74ff.). To this purpose, the embodied energy in a grinding wheel and end mill was determined. The results showed that the embodied energy of the grinding wheel is significantly higher than the embodied energy of the end mill. However, for a given machining task, only one grinding wheel was needed compared with 914 end mills, resulting in a lower overall embodied energy value for the grinding wheel. The study did not consider the environmental impact of the production, use and disposal phases of both tools. Nonetheless, it gives interesting information about the impact of the embodied energy.

Category (II)—Description and Modelling Approaches

Empirical models for grinding processes were investigated by several authors through a technological analysis Knop (1989, pp. 23ff.), Czenkusch (2000, pp. 33ff.), Osterhaus (1994, pp. 24ff.). The models were derived using a nonlinear regression of experimental data and were expressed as a superposition of several exponential expressions (see Eqs. 2–8). This approach was used to derive process models for the technological indicator represented by the grinding force, the tool wear as well as the workpiece surface roughness and hardness. However, the models were limited to the specific grinding process and experimental conditions of testing. Furthermore, the analysis and modelling lacked specific focus on environmental and economic issues.

Category (III)—Description, Modelling and Evaluating Approaches

Munoz and Sheng presented an approach based on physical models to determine the environmental impact of manufacturing processes with geometrically defined cutting edge (Munoz and Sheng 1995, pp. 736ff.). The relationship between the manufacturing process and its environmental impact takes into account material removal mechanics, tool life, scrap production and cutting fluid flow. However, due to lack of life cycle inventory data, no quantitative environmental impact was determined.

With the aim of providing a simple and quick solution in the evaluation of process sustainability, Linke et al. presented an approach based on sustainability indicators (Linke et al. 2013, pp. 556ff.). The sustainability indicator represents a ratio of performance per specific unit of output, by dividing an indicator (e.g. energy demand, cost, productivity) with a normalisation factor (e.g. lifetime, number, weight or units of produced products) (see (Eq. 3.1) (Linke et al. 2013, pp. 556ff.)). The approach, however, lacks detailed information on scale effects and the interdependencies between different indicators, which can result in a biased evaluation.

$$\text{Sustainability indicator} = \frac{\text{Indicator}}{\text{Normalisation factor}} \tag{3.1}$$

In a further approach, Linke et al. presented an indicator to evaluate manufacturing processes according to their sustainability by relating the achieved surface roughness and process energy (Linke et al. 2014, pp. 564ff.). The efficiency indicator η_{sust} is calculated by dividing the delta between measured and targeted quality parameters (Δ Quality parameter) with the delta of measured and targeted energy demand (Δ Energy demand) (see (Eq. 3.2)). The approach focuses on the technological indicator and leaves the environmental and economic indicators out of consideration. Furthermore, no detailed information on targeted values is given, such as how targeted energy demand can be determined.

$$\eta_{sust} = \frac{\Delta\,\text{Quality parameter}}{\Delta\,\text{Energy demand}} \tag{3.2}$$

Based on the extension theory, Yan et al. proposed an approach for the multi-objective evaluation of machining processes (Yan et al. 2014, pp. 1419ff.). Assessment criteria were weighted utilising an entropy weight approach. The realisation of the approach is based on experimental data and empirical models. Due to the application of a weighting system, a qualitative assessment not focusing on the economic indicator is conducted.

In a multi-objective analysis approach, Dettmer investigated the technological, environmental and economic impact of cutting fluids based on grinding experiments as well as the LCA and LCC methods (Dettmer 2006, pp. 66ff.). In the description, as well as in the modelling process, the production, usage and disposal phase of the cutting fluid were all considered. The impact of the cutting fluid on the tool wear and the connected environmental and economic repercussions were not conducted.

Clarens et al. conducted an analysis regarding the technological and economic impact of an environmentally adapted metalworking fluid system (Clarens et al. 2008, pp. 8534ff.). In the technological analysis, the application of mineral oil-based and rapeseed oil-based cutting fluids was compared based on energy demand while considering three different cutting fluid application strategies. Additionally, the environmental impact was evaluated based on an LCA model of both cutting fluids for each application strategy. However, further analyses focusing on the economic impact of the cutting fluid strategy were not conducted.

Category (IV)—Describing, Modelling, Evaluating and Improvement Approaches

The improvement of a machining process was explored in several studies. The improvement approach commonly entails the application of multi-objective improvement algorithms based on empirical process models. The studies differ with regard to the improvement indicators. In the context of cutting processes with geometrically defined cutting edges, studies were conducted by Basu and Sutherland, Liu et al., Kuram et al., Bhusan, Rajemi et al. or Yan and Li, and cutting processes with geometrically undefined cutting edges were investigated by Ilhan et al. Every study used a multi-objective approach to improve several technological indicators, such as process energy, tool wear or surface roughness, by identifying the corresponding process parameters (Ilhan et al. 1992, pp. 435ff.; Liu et al. 2008, pp. 1125ff.; Kuram et al. 2013, pp. 159ff.; Basu and Sutherland 1999, pp. 323ff.; Bhushan 2013, pp. 242ff.; Yan and Li 2013, pp. 462ff.; Rajemi et al. 2010, pp. 1059ff.; Li et al. 2002, pp. 232ff.). The studies differed in terms of the applied weighting, solving algorithm or constraints. However, since only technological indicators were considered, the mentioned improved process parameters represent a "single optimum". This drawback was partially overcome by Wen et al. and Jin et al. by taking into account economic indicators such as production time and production cost in addition to technological values (Wen et al. 1992, pp. 75ff.; Jin et al. 2009, pp. 869ff.).

From an environmental perspective, Srinivasan and Sheng presented a two-step-based approach to improve the manufacturing and planning processes through the consideration of technological and environmental indicators. The first step focuses on the micro-planning of the machining process by improving the selection and application of process parameters, tooling and cutting fluids when machining a feature (e.g. holes, slots, pockets) (Srinivasan and Sheng 1999a, pp. 257ff.). Empirical and physical models are used to describe the cutting process energy demand, process time, cutting fluid loss, tool wear and workpiece quality for turning, milling and drilling. The second step addresses the manufacturing process macro-planning by improving the machining sequence under precedence constraints such as tolerance stack up and accessibility of the features (Srinivasan and Sheng 1999b, pp. 271ff.). In both steps, the improved process scenario is based on a verbal-argumentative evaluation using the criteria process time, process energy and weighted mass. The approach represents a comprehensive procedure to improve the manufacturing and planning processes. However, the approach is limited by an exclusive focus on cutting process energy demand instead of overall energy demand. Furthermore, the approach does not consider further environmental impact due to tool and cutting fluid demand.

3.1.2 System Level

Category (I)—Describing Approaches

In order to describe the impact of cutting fluid and electrical energy demand on technological and economic indicators, Eckebrecht performed analyses on the machine tool and peripheral system level (Eckebrecht 2000, pp. 39). Based on measurements, Eckebrecht calculated/measured the electrical energy demand of a grinding system and investigated the impact of cutting fluid filter type on the overall energy demand (Eckebrecht 2000, pp. 103ff.). The verbal-argumentative description of the grinding process highlighted the need to consider energy and resource flows at the system level; however, the significance is partially constrained by a focus on selected flows.

In an attempt to compare different manufacturing strategies for producing the same product, Morrow et al., Salonitis et al. and Zhao et al. performed investigations exclusively on the machine tool and peripheral system level. In an analysis focusing on the environmental indicator, Morrow et al. compared laser-based, conventional tool and die manufacturing (Morrow et al. 2004, pp. 932ff.). Based on the measured energy demand and material production emissions, an LCA of milling and laser-based direct metal deposition was conducted. Zhao et al. also analysed laser-assisted manufacturing strategies by comparing laser shock peening and laser-assisted turning (Zhao et al. 2010, pp. 1311ff.). A comprehensive LCA of both processes was performed to compare the manufacturing strategies according to their environmental impact. By comparing two manufacturing strategies to accomplish workpiece hardening, Salonitis et al. investigated the grind-hardening process chain and the conventional hardening process chain using an LCA to determine environmental impact (Salonitis et al. 2006 pp. 657ff.). All studies are based on comprehensive life cycle inventories to determine environmental impact. However, the studies lack to consider the economic impact of such processes. Furthermore, the focus on a fixed manufacturing process without any option for the analysis of process parameter variations constrains the transferability to different cases.

Dahmus and Gutowski investigated the relationship of cutting process and non-cutting process-related energy demand by describing the energy demand at the machine tool and peripheral system level (Dahmus and Gutowski 2004, pp. 1ff.). Based on the electrical energy measurement of machine tools with different automation levels, Dahmus and Gutowski showed that the actual needed cutting energy can be comparably lower when considering the total energy demand during material removal (Fig. 3.2) (Dahmus and Gutowski 2004, pp. 3ff.).

Murray et al. described a non-cylindrical grinding process using the LCA method to measure the environmental impact (Murray et al. 2012, pp. 1604 ff.). Focus of the investigation was the description of an existing grinding system and the variation of process parameters; the parameter influence on environmental impact was not considered (Murray et al. 2012, pp. 1609). However, Murray

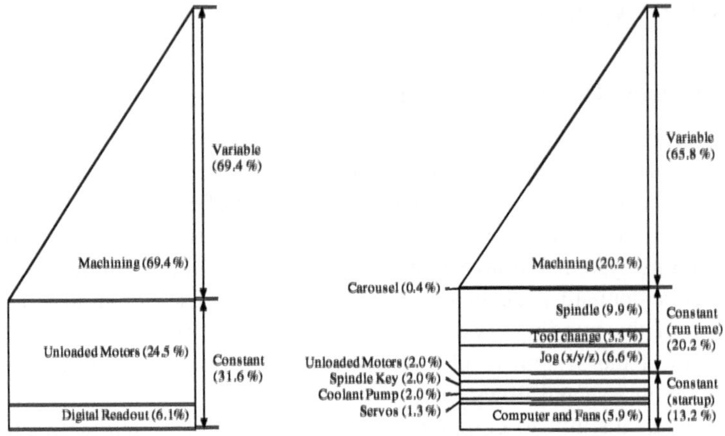

Machining energy use breakdown for a 1985 Bridgeport manual milling machine

Machining energy use breakdown for a 1998 Bridgeport automated milling machine

Fig. 3.2 Effect of automation on the machine tool power distribution (Dahmus and Gutowski 2004, pp. 3)

presents a comprehensive life cycle inventory analysis of the grinding system, considering energy and resource flows.

Category (II)—Describing and Modelling Approaches

The power demand of a machine tool can be described based on the arbitrary operational state of the machine tool components. By measuring the electrical energy demand at each state, the machine tool can be modelled using the proposed discrete event formulation by Dietmair and Verl (2010, pp. 123ff.). The scalable and integrative approach allows a real-time prediction of the energy demand of machine tools for different configurations and different scenarios. Yet, the dependence on measurement data, which are assumed to be constant, limits the applicability of the approach.

To forecast the energy demand of machine tools already in the design phase, Kuhrke pursued an equation-based modelling approach to support decision-making during machine tool development (Kuhrke 2011, pp. 4). The energy demand of each machine tool component was described using physical and empirical models. However, these models are described based on an idealised component operation mode and partially lack to consider friction due to thermal movement or deterioration. Furthermore, the application of empirical models limits the transferability and undermines the aim of the approach, which is to avoid electrical measurements. In addition, the approach focuses solely on the demand of electrical energy.

Category (III)—Describing, Modelling and Evaluating Approaches

Reinhardt pursued an approach to evaluate the resource efficiency of manufacturing process chains based on the derivation of a key performance indicator. The aim was to increase transparency about resource demand efficiency when planning and operating a manufacturing process chain (Reinhardt 2013, pp. 12). Resource efficiency is determined through the inventorisation of energy and resource flows as well as their modelling in input–output balances. Efficiency was then used for the verbal-argumentative evaluation of the environmental impact. The method represents a comprehensive and well-founded approach to determine the resource efficiency of manufacturing process chains. However, the provision of a single indicator can conceal dynamic relationships within the investigated system.

To support decision-making, Schultz proposed an approach for the integrated evaluation of machine tools and peripheral systems based on economic and environmental indicators. For this purpose, the approach used input–output balances (Schultz 2002, pp. 4). The model is based on the inventorisation of energy and material flows as a foundation for the environmental impact assessment and the economic impact assessment (Schultz 2002, pp. 31). Due to a shared database, the approach allows a consistent economic and environmental evaluation of existing and alternative machine tools and peripheral systems. However, the approach lacks the ability to identify machine tool and peripheral system set-ups, which lead to minimal economic and environmental impact.

Thiede pursued a simulation-based approach to evaluate energy demand at the process chain level to support the energy-oriented planning and control of the machine tool and peripheral systems (Thiede 2012, pp. 6ff.). The energetic behaviour of machine tools and technical building services was formulated using empirical and physical models as well as energy measurements and state-related power characterisations. The approach represents a comprehensive method to evaluate process chains based on environmental and economic indicators. However, the application requires detailed knowledge of process chain systems as well as the means to derive the necessary, accurate data for modelling.

To bridge the gap in reaching the ideal energy performance, Zein proposed a performance management concept as well as tools to quantify and evaluate the energy efficiency of machine tools (Zein 2012, pp. 3). Based on electrical measurements, the energy performance limit of machine tools within a specific group of manufacturing processes is determined. The limit is used to derive an energy efficiency ranking of the machine tools (Fig. 3.3). The approach focuses on the energy demand of the machine tool and neglects the impact of the energy demand of supporting systems (e.g. compressed air systems) on the overall efficiency. Furthermore, for the determination of the energy performance limit, a set of comparable and accurate data is crucial to ensure a sufficient degree of accuracy.

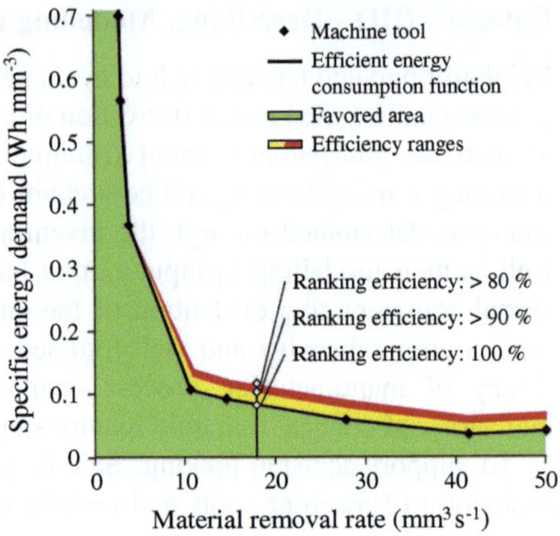

Fig. 3.3 Ranking energy efficiency of machine tools (Zein 2012, pp. 120)

Category (IV)—Describing, Modelling, Evaluating and Improvement Approaches

A systematic four-step-based approach to evaluate and to improve the energy and resource demand at the system level was proposed by Binding (1988, pp. 6ff.). The first step involved the determination of energy and resource flows. In the second step, these flows were analysed with regard to their necessity. The third step focused on the identification of solutions to reduce these flows. The last step involves the evaluation of the achieved improvement. The selection of the common energy and resource flows, however, is based on a cost-related ranking and lacks a focus on environmental indicators.

3.1.3 Process and System Level

Category (I)—Describing Approaches

The description at process and system level can be structured according to the applied manufacturing strategy and technology. The workpiece can either be manufactured using different machine tools within the same group of manufacturing processes or using different groups of manufacturing processes.

The first option was investigated by Behrendt et al. The authors measured the electrical energy demand of nine machine tools designed for machining processes with geometrically defined cutting edges (Behrendt et al. 2012, pp. 43ff.). To compare the machine tools, the authors used a three-step-based measurement approach. In the first step, the idle power is measured, after which the effectiveness of the components is determined (e.g. drives, pumps), and finally a standardised test piece is machined. The results of the study provided a good overview on the influence of machine tool size and configuration on energy demand. The investigation does not seem to take into account further research revolving around environmental and economic indicators.

Kundrák et al. and Kong et al. investigated the second option in order to compare the production of workpieces using different manufacturing processes. Kundrák et al. compared gear manufacturing by using a grinding process or a hard cutting process (Kundrák et al. 2006, pp. 29ff.). Kong et al. compared the manufacturing of a die casting mould based on the application of either a milling process or an electric discharge machining process (Kong et al. 2013, pp. 280ff.). The aim of both studies was the comparison of the manufacturing process according to technological indicators, such as machining time, energy demand or achieved surface roughness. In addition to the technological indicator, Kundrák et al. considered the economic indicator as well; however, both analyses did not take into account further indicators.

Category (II)—Describing and Modelling Approaches

Schiefer developed an approach for the inventorisation of energy and resource flows of cutting processes with geometrically defined cutting edges in support of decision-making in the designing phase of environmentally benign products (Schiefer 2001, pp. 19). For this purpose, the designer defined three subsystems (cutting process, machine tool and supporting system) and described the energy demand in each subsystem based on empirical and physical models (Schiefer 2001, pp. 23ff). Schiefer considered not only energy demand, but also the demand or transformation of raw, auxiliary and operating material flows into waste and emission flows. The successful application of the approach requires the availability of accurate data to determine cutting forces and to describe the machine tools' and supporting systems' energy demand.

A framework for the quantitative thermodynamic analysis of manufacturing processes, machine tool and peripheral system was described by Branham et al. The aim of such framework was to model the demand and transformation of energy and resources (Branham et al. 2008, pp. 1). The modelling approach includes the formulation of the machine tool and peripheral system in mass, energy and entropy balances as well as an exergy-based description of the material removal process efficiency (Branham et al. 2008, pp. 5). The application of the proposed approach facilitates a comprehensive analysis and modelling of manufacturing processes and systems. However, its realisation requires an understanding of the thermodynamic system, as well as the data to entirely formulate the mass, energy and entropy balances.

By investigating the relationship between process parameters, cutting power and the total power demand of the spindle, Draganescu et al. presented a technological approach to model the specific demanded energy in relation to the energy efficiency of the spindle motor (Draganescu et al. 2003, pp. 11.). The spindle motor efficiency η was empirically derived from experimental data and described as a function of multiple process parameters. Efficiency was further used to describe the specific energy consumption (*SEC*) in machining as a function of the cutting power P_c, the material removal rate *MRR* and the efficiency η (see (Eq. 3.3)).

$$\mathrm{SEC} = \frac{P_c}{\eta \cdot \mathrm{MRR}} \tag{3.3}$$

The limitation of the approach is the focus on the spindle motor efficiency instead of the entire machine tool. Furthermore, the approach focuses solely on energy efficiency and does not take into account further technological indicators, such as tool wear and surface roughness as well as environmental and economic indicators.

Based on the theoretical analysis of several cases at the process and system level, Gutowski et al. also presented an empirical process model to determine the specific energy consumption in relation to the material processing rate (Gutowski et al. 2006, pp. 624, 2007, pp. 138). The proposed relationship includes the idle power P_0 of the machine tool, the material processing rate *MRR* and a constant k, which is related to process physics (see (Eq. 3.4)).

$$\mathrm{SEC} = \frac{P_0}{\mathrm{MRR}} + k \tag{3.4}$$

Due to its empirical nature, the equation is only valid for a specific manufacturing process. Furthermore, the approach lacks an explanation of the constant k related to the physics of the process as well as an initial validation. The approach proposed by Gutowski et al. was further investigated by Li and Kara, Diaz-Elsayed as well as Helu, in order to overcome its limitations. Li and Kara validated the inverse regression model proposed by Gutowski et al. on the basis of experimental data. Furthermore, the authors conducted a modification of the idle power P_0 and the constant k by using the coefficients c_0 and c_1. The material removal rate remained the same (see (Eq. 3.5)) (Li and Kara 2011, pp. 1640).

$$\mathrm{SEC} = c_0 + \frac{c_1}{\mathrm{MRR}} \tag{3.5}$$

The adapted model showed a high level of statistical accuracy in quantifying specific energy consumption in relation to the material removal rate for turning, milling and grinding as well as injection moulding processes (Li 2012, pp. 44ff.). Furthermore, Li proposed a qualitative explanation for the coefficients c_0 and c_1. Coefficient c_0 can be written as a function of the spindle drive motor characteristics, workpiece material, tool geometry and cutting parameters, while c_1 can be described as a function of the fixed power demand, motor and transmission efficiency (Li 2012, pp. 113ff.). Diaz-Elsayed as well as Helu used the approach by Li and Kara and proposed as well an approach to quantify the specific energy consumption in relation to the material removal rate of different machining processes with geometrically defined cutting edges (Diaz-Elsayed 2013, pp. 42ff.; Helu 2013, pp. 43ff.). Furthermore, all three authors considered not only energy efficiency as a main technological indicator, but also surface roughness. Additionally, Li et al. incorporated environmental indicators in the case of a grinding process (Li et al. 2012, pp. 59ff.), while Helu et al. focused on economic indicators (Helu et al. 2012, pp. 55ff.).

Category (III)—Describing, Modelling and Evaluating Approaches

In order to evaluate the process and system level, Wolfram proposed the embodied energy approach (Wolfram 1986, pp. 11ff.). In this approach, the primary energy demands of the processed material, the tool as well as the operation of the machine tool and peripheral system during the cutting process were accumulated on the basis of electrical measurements. The approach allowed the comparison of different manufacturing processes and systems. In addition, the approach fosters the ability to identify improvement potential, due to its holistic consideration of energy demands. The dependence on measurement data, however, limits the transferability of the approach.

Creyts and Carey proposed the application of extended exergy analysis to evaluate the environmental performance of manufacturing processes (Creyts and Carey 1999, pp. 247). The approach starts with modelling and calculating the exergy connected to the system's input and output flows. The approach also takes into account the total exergy used to separate the mixed output streams as well the exergy to return the output streams to the environment (Creyts and Carey 1999, pp. 254). The model is then used to evaluate process and system parameter modifications with regard to the related exergy demand as a basis for a verbal-argumentative evaluation of the environmental performance. The realisation of this approach facilitates a comprehensive system description. However, it also requires an understanding of the thermodynamical system as well as the availability of the necessary input and output data.

The modelling of the process and system levels based on input–output balances was applied by Choi et al., Xue et al. and Jiang et al. The focus of Choi et al. was the application of input–output balances as a basis for a life cycle assessment of manufacturing processes fostering decision support for alternative scenarios during the product design phase (Choi et al. 1997, pp. 231). A verbal-argumentative evaluation was performed based on the analyses of manufacturing process-related energy demand, noise emissions, produced solid waste and incurred wastewater. However, the approach neglects further energy and resource flows caused by the demand and transformation of tools, cutting fluid or machine tool. Furthermore, the approach does not take into account the connected economic and technological impact. Xue et al. pursued the goal of modelling the connection between unit-level changes and broader system behaviour (Xue et al. 2007, pp. 1349). For this purpose, input and output flows of the systems were analysed and on this basis, input–output balances were described. The balances, as seen in the approach of Choi et al., served as a verbal-argumentative evaluation of the environmental impacts.

Given the aim of providing a method for an aggregated system evaluation, the approach lacks an in-depth evaluation of the dynamics occurring at the process system level. Jiang et al. applied the input–output balance to evaluate the environmental impact of manufacturing processes and process plans. The proposed approach utilised a three-step procedure including the determination of process inventories, quantifying environmental impacts and evaluating the environmental performance (Jiang, et al. 2012, pp. 784). The approach is based on measurement data and does not consider the process dynamics. Furthermore, the proposed approach applies a verbal-argumentative evaluation of the environmental impact.

Anderberg proposed an extended cost model for machining processes, which includes the economic indicator. He included the costs of carbon dioxide emissions linked with energy demand and the direct process energy costs, in addition to the traditional costs for machining hours, labour, energy, tools, etc. (Anderberg 2012, pp. 33ff.). The results of a case study indicated a comparably low share of costs due to energy-related carbon dioxide emissions and direct process energy in relation to the total machining costs (Anderberg et al. 2009, pp. 531ff.). A drawback of the model is the focus on energy-related carbon dioxide emissions only and its disregard of further carbon dioxide emissions resulting from tool wear or cutting fluid demands.

Due to the lack of available and reliable life cycle inventory data of metal removal processes, Kellens proposed a methodology for the systematic analysis and evaluation of the manufacturing unit process life cycle inventory. The aim of the methodology was the reduction of energy-related environmental impact of discrete part manufacturing processes (Kellens 2013, pp. 1ff.). The approach includes a screening step to estimate the fixed and process-dependent energy demand based on functional and physical models. The following step entails an in-depth analysis, including time and power studies for the detailed determination of machine tool and peripheral system energy demand (Kellens et al. 2012, pp. 69ff). The approach provides a structured procedure and has been validated for bending and milling (Devoldere et al. 2007, pp. 311ff.), laser cutting, selective laser sintering as well as electrical discharge machining processes (Kellens 2013, pp. 47ff.). However, the amount of time needed for data collection (one to two weeks) limits a fast realisation (Kellens et al. 2012, pp. 74), and one that does not take into account demanded resources.

To support decision-making during manufacturing process planning, Narita et al. proposed an evaluation system used to determine the environmental impact of machining strategies generated by computer-aided manufacturing software (Narita et al. 2006, pp. 1188ff., 2008, pp. 65ff.). The approach comprises an equation-based description of relationships between process parameters and demanded or transformed energy and resource flows within the machine tool and peripheral system. The equivalent CO_2 emission intensities caused by the flows, as well as the environmental impact, were calculated based on machine tool motions and machining operations (Narita et al. 2006, pp. 1189). The realisation of the approach depends on data availability of demanded and transformed energy and resource flows as well as the availability of CO_2 emission intensities linked to environmental impact.

Furthermore, while focusing on environmental indicators, the approach neglects technological and economic ones as further dimensions in the support of decision-making during manufacturing process planning.

Schlosser proposed an equation-based approach in the evaluation of machine tools and peripheral systems in relation to technological, environmental and economic indicators (Schlosser 2013, pp. 1ff.). To this end, physical and empirical process and system models were used to describe demanded and transformed energy and resource flows during milling, turning and drilling processes in a machine tool. Furthermore, the approach includes the energy demand of manufacturing process supporting systems as well as environment conditioning systems, e.g. compressed air generation, shop floor ventilation and illumination, and technical heating and cooling.

Shin introduced a further approach for an integrated evaluation of manufacturing processes by considering the relationship between process parameters and the machine tools' energy and resource demand to derive an integrated Green Productivity Index (GPI) (Shin 2009, pp. 42ff.). The focus of the approach is its application as a planning tool for the selection of process parameters prior to the manufacturing process. Based on the measurement parameters' machining time, roughness, tool using time, tool wear, energy demand, chip mass, air emission and cutting fluid loss, the GPI was determined using the performance and sub-performance criteria (Fig. 3.4).

An expert knowledge-based weighting method was used to determine the relationship between the measurement parameters and the GPI. The weighting method used a relational matrix between measurement parameters sub-performance criteria, performance criteria and green productivity level (Shin 2009, pp. 54). To reduce the effort in conducting the approach, measurement parameters are partially described using empirical models (e.g. energy demand, tool wear) or automatically measured

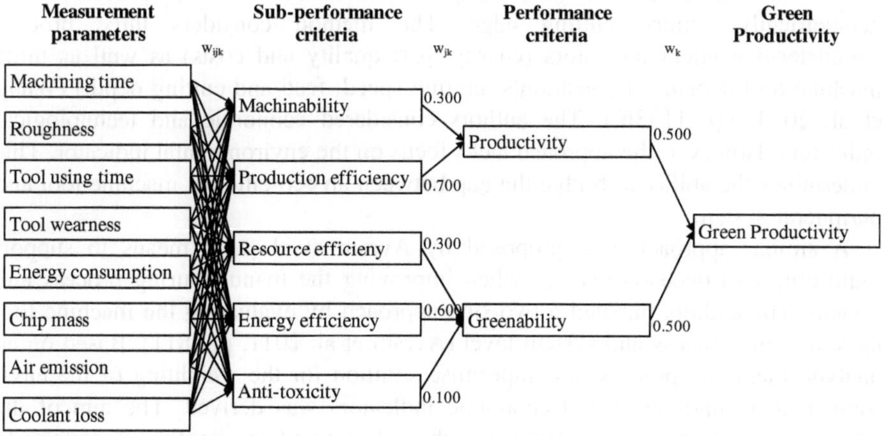

Fig. 3.4 Exemplary tree diagram between measurement parameters and green productivity (Shin 2009, pp. 55)

while performing the manufacturing process (e.g. air pollutant, coolant loss) (Shin 2009, pp. 57). The objectives of the approach are constrained because of the application of an expert weighting method. A further constraint is the dependence of the approach on empirical models and measurement data that can only be derived during the process and its limits its applicability.

To enable an integrated sustainability evaluation of process and machine tool and peripheral systems, Zhang and Haapala proposed an approach that considers economic, environmental and social indicators (Zhang and Haapala 2014, pp. 1ff.). Based on the Preference Ranking Organization Method for Enrichment of Evaluations (PROMETHEE) method, the approach evaluates different production scenarios and supports multi-objective decision-making that takes into account the relevant energy and resource flows. The approach provides an integrated method to evaluate the sustainability of manufacturing process scenarios based on measured data. However, it overlooks empirical or physical models to flexibly represent changed requirements in the manufacturing process scenario.

Category (IV)—Describing, Modelling, Evaluating and Improvement Approaches

Focusing on energy- and resource-related restrictions, Larek proposed a multi-objective simulation approach to improve a process chain consisting of the turning, hardening and grinding manufacturing steps (Larek 2012, pp. 1ff.). Each manufacturing process was empirically modelled on the basis of process parameters-related energy and resource demand of energies, cutting fluids and tools. The approach allowed the determination of the process parameters leading to minimal energy and resource demand of the process chain. Yet, it lacks the consideration of the environmental and economic indicators.

A further approach based on the application of empirical models to support machine tool and peripheral system improvement was proposed by Wang et al. The authors formulated a multi-objective improvement model for cutting processes with geometrically defined cutting edge. The method considers three process parameter-dependent indicators (energy, part quality and costs) as well as three machine tool-dependent constraints (cutting speed, feed and cutting depth) (Wang et al. 2014, pp. 1133ff.). The authors considered economic and technological indicators. However, the approach lacks focus on the environmental indicator. This undermines the ability to bridge the gap between an eco-efficient machine tool and peripheral system.

A similar approach was proposed by Avram et al. as a means to support multi-objective decision-making when improving the manufacturing process and system. The authors pursued a two-step approach by evaluating the machine tool system at the process and system level (Avram et al. 2011, pp. 811). Based on an analytic hierarchy process, a compromise solution for the weighting of the environmental, technological and economic indicators was derived. The aim of the improvement at the process level was the selection of the "optimal" strategy to machine a specific feature, while the system level focused on the selection of the most appropriate machine tool system for a use phase scenario (Avram et al. 2011, pp. 819). Measurements on each level provided the necessary database. However, the dependence of this approach on measurement data constrains its applicability.

3.2 Comparative Evaluation of Current Research Approaches

The approaches described in the previous section were evaluated in a comparative review. For this purpose, from a total number of 70 approaches, only 30 investigations were selected. The selection excludes investigations, which focus only on the description of the process level and/or system level. Furthermore, approaches focusing solely on the technological indicator and neglecting the environmental and/or economic indicators were not considered. This reduction is due to the overall focus of this approach on describing, modelling, evaluating and improving the eco-efficiency at the process and system level. Based on the deduction of comparison criteria in Sect. 3.2.1, a comparative evaluation of the approaches follows in Sect. 3.2.2 and the need for further research is presented in Sect. 3.2.3.

3.2.1 Deduction of Comparison Criteria

Overall, 15 sub-criteria are deduced and classified into six main categories. Each sub-criterion is structured into either two or four characteristic gradations in order to present the degree of criteria fulfilment. If an approach comprises one or more of these characteristics, then a quadrant ((◔)) is added. These main categories and sub-categories, as well as the extent to which they are fulfilled, are described in more detail in this section. The main and sub-criteria are presented in Table 3.1.

The main criterion **"general"** is composed of three sub-criteria: *scope*, *level* and *assessment*. The purpose of the general criterion is to give an overall overview of the classification of the approach. The sub-criterion *scope* is structured according to the fulfilment of the following characteristics assessing the approach focus: description, modelling, evaluation and improvement (based on Sect. 3.1 and Fig. 3.1). The sub-criterion *level* comprises the following hierarchical levels: chip formation, process, machine system and manufacturing system. In order to evaluate if an approach has an economic, an environmental, a technological and/or an eco-efficiency orientation, the sub-criterion *assessment* is necessary.

Within the main criterion **"description"**, the sub-criteria *completeness* and *systemdynamics* are considered. The sub-criterion *completeness* represents the degree of considered energy and resource flows. The characterisation graduation can tell whether an approach considers just one single energy or resource flow or

Table 3.1 Criteria for evaluation of research approaches

Criteria	Characteristics of fulfilment				
	○	+ ◔	+ ◑	+ ◕	+ ●
General					
Scope	Not considered	Description	Modelling	Evaluation	Improvement
Level	Not considered	Chip formation	Process	Machine system	Manufacturing system
Assessment	Not considered	Economic	Environmental	Technological	Eco-efficiency
Description					
Completeness	Not considered	Single energy flow	Single resource flow	Multiple energy flows	Multiple resource flows
System dynamics	Single value	Single parameter	Set of single parameters	Multiple parameters	Set of multiple parameters
Modelling					
Approach	Not considered	Empirical of single value	Empirical of multiple value	Physical of single value	Physical of multiple value
Application	Not considered	Theoretical /conceptional equations	Calculation of measured value	Measurement based forecasting	Parameter-based forecasting
Evaluation					
Economic	Not considered	Verbal-argumentative	Conversion to cost	Simplified cost model	Complete cost model
Environmental	Not considered	Verbal-argumentative	Conversion to one impact factor	Simplified LCA	Full LCA
Technological	Not considered	Verbal-argumentative	Process results and condition	Resource flows	Energy flows

(continued)

Table 3.1 (continued)

Criteria	Characteristics of fulfilment				
	○	+ ◔	+ ◔	+ ◔	+ ◔
Improvement					
General	Not considered	Addressed and possible, yet not conducted		Improvement studies are conducted	
Focus	Not considered	Single-objective improvement		Multi-objective improvement	
Implementation					
Knowledge level	Expert knowledge	No specific knowledge necessary	No general knowledge necessary	No preliminary knowledge necessary	Ready to use
Effort	Complex to solve	No high effort necessary	No medium effort necessary	Low effort necessary	Ready to use
Support	Not provided	Support to foster understanding	Support for data acquisition	Support for decision-making	Support for systematic improvement

whether multiple energy and resource flows are used. The sub-criterion *system dynamics* includes the number of considered process and system parameters. This entails a differentiation if the approach uses a single parameter (e.g. $a_e = 1$ µm), a set of single parameters' (e.g. $a_e = 1$ µm, ..., $v_c = 60\,\frac{\text{m}}{\text{s}}$), multiple parameters' (e.g. $a_e = 1 \ldots n$ µm) and multiple parameter sets (e.g. a set of $a_e = 1, \ldots, n$ µm, ..., $v_c = 60, \ldots, n$ m/s).

The approaches are categorised within the main criterion "**modelling**" according to the sub-criteria *approach* and *application*. The sub-criterion *approach* is differentiated in accordance with the chosen modelling characteristics (empirical or physical) used to describe aspects (single or multiple). The sub-criterion *application* characterises whether the approach focuses on the provision of theoretical/conceptional equations without proof, on calculation of measured values using empirical or physical formulations, on the forecasting of values based on measurements or the forecasting of values based on parameter-dependent models.

The main criterion "**evaluation**" is structured into the three sub-criteria *economic*, *environmental* and *technological*, and characterises the detail and focus of the assessment procedure. The sub-criterion *economic* differentiates based on whether the costs are considered verbal-argumentatively, on conversion to a cost value, in a simplified model or in a complete model. A similar distinction can be made with regard to the *environmental* sub-criterion. Instead of using costs, the environmental impact of single or multiple flows can be represented using single impact categories (e.g. kg CO_2-eq.), a simplified LCA or a full LCA. The *technological* sub-criterion differentiates on whether the impact is evaluated verbal-argumentatively, by the presentation of process results (e.g. R_a) or conditions (e.g. F_t, F_n), resource flows or energy flows.

The main criterion "**improvement**" includes the two sub-criteria *general* and *focus*. It evaluates the improvement focus of the approaches. The sub-criterion *general* differentiates based on whether the improvement is addressed and possible, yet not conducted, or whether an improvement study is conducted. The sub-criterion *focus* divides the approaches into a single- or multi-objective improvement.

The last main criterion "**implementation**" includes the three sub-criteria *knowledge level*, *effort* and *support*. This main criterion focuses on the assistance and possibility to apply the approach. The *knowledge level* sub-criterion evaluates the approaches by differentiating on the basis of whether expert knowledge, no general system knowledge, no specific system knowledge and no previous knowledge are necessary or a ready-to-use situation already exists. The sub-criterion *effort* evaluates the necessary efforts with regard to time, cost and material. The sub-criterion differentiates between a complex situation to solve, no high effort necessary, no medium effort necessary, low effort necessary and a ready-to-use situation. The sub-criterion *support* characterises the type of support needed when the user applies the approach. The sub-criterion is divided between no support, support to foster understanding, support for data acquisition, support for decision-making and support for systematic improvement.

3.2.2 Comparative Overview

The research approaches presented in the comparative overview in Sect. 3.1 are linked by the presence of the 15 sub-criteria listed in the previous section. The results are presented in Tables 3.2 and 3.3. The tables feature a matrix form. The revised approaches are presented on the left side, in connection with the sub-criteria at the top. The fulfilment of a sub-criteria characteristic is indicated with a black quadrant (◔).

The first main criterion, "general", represents the classification of the approaches as shown in Fig. 3.1. The fulfilment of the scope sub-criterion indicates a high proportion (circa 60 %) of approaches focusing on the description, modelling and evaluation of the investigated process and/or system. Almost a quarter of the considered research approaches comprises description, modelling, evaluation and improvement steps. Only a minor proportion focuses on description and modelling. As far as the sub-criterion "level" is concerned, the considered approaches focus predominantly on the manufacturing process (circa 30 %) and on the combination of manufacturing process and system (circa 50 %). Only a minority also considers chip formation in addition to the process and system levels. The combination of technological and environmental assessment (circa 53 %) prevails in the case of the *assessment* sub-criterion. Only a small fraction takes into account the economic assessment. No approach makes use of all three categories simultaneously to determine eco-efficiency.

The detailed analysis of the "description" main criterion shows a high degree of fulfilment with regards to completeness and system dynamics. Approaches based on a thermodynamic perspective provide a high degree of fulfilment due to their comprehensive understanding of the system (e.g. Branham et al., Creyts and Carey or Munoz and Sheng). Overall, circa 30 % of the considered approaches fulfil all of the characteristics. This high percentage is followed by a further 37 % of the considered investigations only lacking the integration of multiple energy flows to achieve a complete fulfilment of the defined characteristics. Other approaches, however, focus solely on a single system indicator (e.g. energy). In the case of sub-criterion "system dynamics", all characteristics (circa 46 %) were present. In a considerable number of approaches, however, the investigation rests on a single parameter (circa 30 %) or a single set of parameters (circa 20 %).

The main criterion, "modelling", shows a comparably higher degree of fulfilled criteria. Empirical modelling in particular (circa 47 % of the considered approaches) prevails over physical modelling (circa 20 % of the considered investigations). A combination of empirical and physical modelling is applied in certain cases (circa 33 % of the total). Regarding the application focus of the modelling procedure, the fulfilment of all characteristics is conducted by only 23 % of the approaches. The only missing element in around 36 % of the considered approaches is parameter-based forecasting.

As far as "evaluation" is concerned, the investigated approaches only partially fulfil the criteria. In case of the economic evaluations, the verbal-argumentative

Table 3.2 Comparative overview on the state of research

	General			Description		Modelling		Evaluation			Improvement		Implementation		
	Scope	Level	Assessment	Completeness	System dynamics	Approach	Application	Economic	Environmental	Technological	General	Focus	Knowledge level	Effort	Support
Anderberg, 2012	◕	◕	⊕	◑	◕	◔	◔	◕	○	◔	○	○	◑	◑	◔
Avram et al., 2011	●	◒	◒	◑	◑	◔	◔	◔	◔	◔	○	○	◑	◑	⊕
Branham et al., 2008	◑	◒	◒	●	●	●	◓	○	○	○	○	○	○	○	◔
Choi et al., 1997	◕	◒	◒	◓	◔	◔	◔	○	◔	◕	○	○	◔	◔	⊕
Clarens et al., 2008	◕	◒	◒	◕	◔	◔	◑	○	●	◑	○	○	◑	◑	◕
Creyts & Carey, 1999	◕	◒	◒	●	●	●	◓	○	◔	◕	○	○	○	○	⊕
Diaz-Elsayed, 2013	◑	◒	◒	◔	◔	◔	◕	○	○	○	○	○	◑	◑	◑
Dettmer, 2006	◕	◔	◕	⊕	◔	◔	◑	◑	●	◑	○	○	◑	◑	◕
Helu, 2013	◑	◒	◕	◔	◔	◔	◕	○	○	○	○	○	◑	◑	◑
Jiang et al., 2012	◕	◒	◒	●	◔	◔	◔	◔	◔	◔	○	○	◔	◔	⊕
Jin et al., 2009	●	◔	⊕	●	●	◐	○	●	◑	◔	●	●	◔	◔	●
Kellens, 2013	◕	◒	◒	●	●	◑	◕	◔	●	◕	○	○	◑	◑	◕
Larek, 2012	●	◒	⊕	◕	●	◑	◕	◑	○	●	●	●	◔	◔	●
Li, 2012	◑	◒	◕	◕	◔	◔	◕	○	◕	●	○	○	◑	◑	⊕
Linke et al., 2013	◕	◔	◒	◕	◔	◔	◔	◕	◑	◔	○	○	◑	◑	⊕

Table 3.3 Comparative overview on the state of research

	General			Description		Modelling		Evaluation			Improvement		Implementation		
	Scope	Level	Assessment	Completeness	System dynamics	Approach	Application	Economic	Environmental	Technological	General	Focus	Knowledge level	Effort	Support
Munoz & Sheng, 1995	◕	¼ (lower right)	◒	●	●	●	●	◔	◔	½ (upper right + lower left)	○	○	◔	◔	½ (upper right + lower left)
Narita et al., 2006	◕	◒	◒	¾ (lower left open)	●	◐	●	◑	◑	◔	○	○	◔	◔	◕
Reinhardt, 2013	◕	¼ (lower left)	◒	¾ (lower left open)	◑	◐	◔	◔	◔	¾ (lower right open)	○	○	◔	◔	◕
Schiefer, 2001	◑	◕	◒	●	◑	◑	◕	○	○	¾ (lower right open)	○	○	◔	◔	½ (upper right + lower left)
Schlosser, 2013	◕	◕	◕	●	●	●	◕	●	◕	●	○	○	◔	◔	●
Schultz, 2002	◕	¼ (lower left)	¾ (lower left open)	●	◑	◔	◕	●	◕	○	○	○	◑	◑	½ (upper right + lower left)
Shin, 2009	◕	◕	◕	¾ (lower left open)	◑	◑	◕	◔	◔	●	○	○	◑	◑	½ (upper right + lower left)
Srinivasan & Shen, 1999a	●	¼ (lower right)	◒	¾ (lower left open)	●	●	●	◔	◔	¾ (lower right open)	●	◑	◔	◔	●
Srinivasan & Shen, 1999b	●	¼ (lower right)	◒	¾ (lower left open)	●	●	●	◔	◔	¾ (lower right open)	●	◑	◔	◔	●
Thiede, 2012	◕	¼ (lower right)	◕	½ (upper right + lower left)	●	◐	●	◕	◑	¾ (lower right open)	○	○	◔	◔	●
Wang et al., 2014	●	◒	½ (upper right + lower left)	●	●	●	◕	◔	◔	●	●	●	◔	◔	●
Wen et al., 1992	●	¼ (lower right)	½ (upper right + lower left)	½ (upper right + lower left)	●	●	●	◕	○	●	●	●	◔	◔	●
Xueet al., 2007	◕	◒	◒	½ (upper right + lower left)	◔	◐	½ (upper right + lower left)	◔	◔	¾ (lower right open)	○	○	◔	◔	◑
Yan et al., 2014	◕	¼ (lower right)	¼ (lower right)	¾ (lower left open)	●	●	●	◔	◔	◔	○	○	◔	◔	½ (upper right + lower left)
Zhang & Haapala, 2014	◕	◒	◕	¾ (lower left open)	◑	◐	◕	●	●	◔	○	○	◔	◔	◕

evaluation is predominant (circa 37 %). Only 13 % of the considered approaches fulfil all the characteristics, and around 26 % do not take into account economic impact at all. The environmental evaluation is characterised by the predominance of the verbal-argumentative evaluation (circa 37 %). Only about 13 % of the considered approaches fulfil all the characteristics. However, in the case of technological evaluation, only 13 % do not conduct any assessment at all. Additionally, about 27 % of the approaches perform a verbal-argumentative evaluation only. The same percentage was found in the additional evaluation of energy and resource flows.

As far as "improvement" is concerned, only six out of 30 approaches contemplated this procedure, and they fulfil most of the remaining main criteria compared to the other approaches. Four of the aforementioned six approaches apply a multi-objective improvement algorithm.

The "implementation" main criterion indicates that a majority of the approaches focus on the provision of decision-making support and systematic improvement. Only a minor share focuses on the support of data acquisition and understanding. With regard to the effort and knowledge level required to pursue the implementation, the approaches show an inferior or at least a medium degree of fulfilment, due to the necessity of specific or general knowledge as well as a comparably high effort for preparation and realisation.

3.2.3 Derivation of Research Demand

The comparative overview from the previous section indicates that none of the listed approaches has a high degree of criteria fulfilment, nor is there a balanced fulfilment of all the criteria. Rather, the focus of the approaches is on description, simple modelling and technological evaluation of either process or system, but not both of them. The approaches only partially consider the interrelations between chip formation, process and machine tool as well as detailed economic and environmental evaluations to determine eco-efficiency or improvement procedures. The presented approaches predominantly focus on machining processes with geometrically defined cutting edges, while processes with geometrically undefined cutting edges are considered insufficient. However, the latter processes in particular have a high significance in the machining of materials in terms of quality, time, energy and resource demand (see Sect. 2.1).

This analysis and the comparative overview in Sect. 3.2.2 indicate the need for additional work in many aspects. The **coherent description, modelling, evaluation and improvement** of a manufacturing process and system are only reflected in a minor share of investigations. Yet, a coherent procedure allows support for the planning of a new or the improving of an existing process and system in an integrated way. Accordingly, current approaches partially constrain themselves. Many approaches focus on the provision of a procedure which considers the peculiarities of either **process** or **system level**. Only a minor share considers the interactions between the process and system levels. Additionally, the main focus is on machining processes with geometrically defined cutting edges. Machining processes with geometrically undefined cutting edges, however, have a major economic, technological and environmental influence.

Many approaches already have a sufficient level of **description** completeness and put the necessary emphasis on system dynamics. However, the comparative analysis also showed that the full range of relevant energy and resource flows, which can be described as functions of various parameters, has not been considered so far. The **modelling** of the relevant energy and/or resource flows is predominantly achieved using empirical models. It can be assumed that the reason for empirical modelling lies within its simplicity or its fast realisation. In order to apply a solution which is independent from a specific case, physical models need to be applied. If the application of physical models alone is not possible, then a combination of physical and empirical models can be an option. Many approaches provide the means to achieve a verbal-argumentative **evaluation** of the investigated process and/or system using technological and economic indicators or technological and environmental indicators. However, the combined application of technological, economic and environmental indicators has been rarely considered so far. The simultaneous use of all three perspectives allows the examination and minimisation of goal conflicts among the indicators. Only a minor share of the considered approaches entails the application of an **improvement** step. It can be assumed that the complexity and effort in conducting an improvement step constrain its realisation. However, the improvement of a process and system allows the structured and substantiated identification of potentials in the reduction of the environmental impact or in the increase of economic and technological value.

Many studies provide support to decision-making in the realisation of the proposed approach during an **application cycle**. However, the realisation of these investigations often requires detailed knowledge of the investigated process and/or system. Furthermore, a comparably higher effort for preparation and realisation needs to be taken into account in such instances.

Accordingly, the need for a conceptual and coherent procedure to determine the eco-efficiency of grinding processes and systems becomes evident. Such a procedure includes a detailed description of the relevant energy and resource flows based on parameters and properties. This description forms the foundation of mixed physical–empirical modelling to allow a detailed and accurate formulation of the relevant energy and material flows of the grinding process and system. Based on these physical–empirical models, an integrated evaluation of the technological, economic and environmental indicators is achievable. In order to consider and minimise goal conflicts, an improvement step is necessary to identify potential to reduce the environmental impact or to increase the economic and technological value. The procedure needs a high degree of flexibility to assess the influence of varying parameters and configurations during the planning of new or the improvement of existing grinding processes and systems.

References

Anderberg, S. (2012): Methods for improving performance of process planning for CNC machining. Ph.D Thesis, Chalmers University of Technology, Göteborg, Sweden.

Anderberg, S., Kara, S., Beno, T. (2009): Impact of energy efficiency on computer numerically controlled machining. Proceedings of The Institution of Mechanical Engineers Part B-journal of Engineering Manufacture, 224/4:531-541.

Aurich, J.C., Carrella, M., Steffens, M. (2012): Evaluation of Abrasive Processes and Machines with Respect to Energy Efficiency. Proceedings of the 19th CIRP Conference on Life Cycle Engineering, University of California at Berkeley, Berkeley, USA, May 23-25, 329-333.

Aurich, J.C., Linke, B., Hauschild, M., Carella, M., Kirsch, B. (2013): Sustainability of abrasive processes. CIRP Annals - Manufacturing Technology, 62/2:653-672.

Avram, O., Stroud, I., Xirouchakis, P. (2011): A multi-criteria decision method for sustainability assessment of the use phase of machine tool systems. The International Journal of Advanced Manufacturing Technology, 53/5-8:811-828.

Basu, S., Sutherland, J.W. (1999): Multi-objective decision making in environmentally conscious manufacturing. Presented at the 6th CIRP International Seminar on Life Cycle Engineering, 323-331.

Behrendt, T., Zein, A., Min, S. (2012): Development of an energy consumption monitoring procedure for machine tools. CIRP Annals - Manufacturing Technology, 61/1:43-46.

Bhushan, R.K. (2013): Optimization of cutting parameters for minimizing power consumption and maximizing tool life during machining of Al alloy SiC particle composites. Journal of Cleaner Production, 39:242-254.

Binding, H.J. (1988): Grundlagen zur systematischen Reduzierung des Energie- und Materialeinsatzes. Dr.-Ing. Dissertation, RWTH Aachen, Aachen, Germany.

Branham, M., Gutowski, T., Jones, A., Sekulic, D. (2008): A Thermodynamic Framework for Analyzing and Improving Manufacturing Processes. 16th IEEE International Symposium on Electronics and the Environment, San Francisco, USA, May 19-20, 1-6.

Choi, A.C.K., Kaebernick, H., Lai, W.H. (1997): Manufacturing processes modelling for environmental impact assessment. Journal of Materials Processing Technology, 70/1-3:231-238.

Clarens, A.F., Zimmermann, J.B., Keoleian, G.A., Hayes, K.F., Skerlos, S.J. (2008): Comparison of Life Cycle Emissions and Energy Consumption for Environmentally Adapted Metalworking Fluid Systems. Environmental Science & Technology, 42/22:8534-8540.

Creyts, J.C., Carey, V.P. (1999): Use of extended energy analysis to evaluate the environmental performance of machining processes. Proceedings of the Institution of Mechanical Engineers, Part E: Journal of Process Mechanical Engineering, 213/4:247-264.

Czenkusch, C. (2000): Technologische Untersuchungen und Prozessmodelle zum Rundschleifen. Dr.-Ing. Dissertation, Universität Hannover, Fortschrittsbericht, VDI Verlag, VDI-Reihe 2/530.

Dahmus, J.B., Gutowski, T.G. (2004): An Environmental Analysis of Machining. Proceedings of ASME International Mechanical Engineering Congress and R&D Exposition, 13-19.

Denkena, B., Reichstein, M., Kramer, N., Jacobsen, J., Jung., M. (2005): Eco- and energy-efficient grinding processes. Key Engineering Materials, 291-292:39-44.

Dettmer, T. (2006): Nichtwassermischbare Kühlschmierstoffe auf Basis nachwachsender Rohstoffe. Dr.-Ing. Dissertation, Technische Universität Braunschweig, Vulkan Verlag, Essen. Germany.

Devoldere, T., Dewulf, W., Deprez, W., Willems, W., Duflou, J.R. (2007): Improvement potential for energy consumption in discrete part production machines. Proceedings of the 14th CIRP Conference on Life Cycle Engineering, g, Waseda University, Tokyo, Japan, June 11–13, 311-316.

Diaz-Elsayed, N. (2013): Development of Energy Models for Production Systems and Processes to Inform Environmentally Benign Decision-Making. Ph.D Thesis, University of California, Berkeley, USA.

Dietmair, A., Verl, A. (2010): A generic energy consumption model for decision making and energy efficiency optimisation in manufacturing. International Journal of Sustainable Engineering, 2/2:123-133.

Draganescu, F., Gheorghe, M., Doicin, C.V. (2003): Models of machine tool efficiency and specific consumed energy. Journal of Materials Processing Technology, 141/1:9-15.

Eckebrecht, J. (2000): Umweltverträgliche Gestaltung von spanenden Fertigungsprozessen. Dr.-Ing. Dissertation, Universität Bremen, Shaker Verlag, Aachen, Germany.

Frăţilă, D. (2009): Evaluation of near dry machining effects on gear milling process. Journal of Cleaner Production, 17/9:839-845.

Frăţilă, D. (2010): Macro-level environmental comparison of near-dry machining and flood machining. Journal of Cleaner Production, 18/10-11:1031-1039.

Garcia, E., Pombo, I., Sanchez, J.A., Ortega, N., Izquierdo, B., Plaza, S., Marquinez, J.I., Heinzel, C., Mourek, D. (2013): Reduction of oil and gas consumption in grinding technology using high pour-point lubricants. Journal of Cleaner Production, 51:99-108.

Gutowski, T.G., Dahmus, J.B., Thiriez, A. (2006): Electrical energy requirements for manufacturing processes. The 13th CIRP International Conference on Life Cycle Engineering, Katholieke Universiteit Leuven, Belgium, May 31 - June 2, 623-627.

Gutowski, T.G., Dahmus, J.B., Thiriez, A., Branham, M., Jones, A. (2007): A thermodynamic characterization of manufacturing processes. IEEE International Symposium on Electronics and the Environment, 137-142.

Helu, M. (2013): Leveraging Manufacturing Precision to Reduce Product Life Cycle Environmental Impacts. Ph.D Thesis, University of California, Berkeley, USA.

Helu, M., Behmann, B., Meier, H., Dornfeld, D., Lanza, G. (2012): Impact of green machining strategies on achieved surface quality. CIRP Annals - Manufacturing Technology, 61/1:55-58.

Ilhan, R.E., Sathyanarayanan, G., Storer, R.H., Liao, T.W. (1992): Off-line multiresponse optimization of electrochemical surface grinding by a multi-objective programming method. International Journal of Machine Tools and Manufacture, 32/2:435-451.

Jiang, Z., Zhang, H., Sutherland, J.W. (2012): Development of an environmental performance assessment method for manufacturing process plans. The International Journal of Advanced Manufacturing Technology, 58/5-8:783-790.

Jin, K., Zhang, H.C., Balasubramaniam, P., Nage, S. (2009): A multiple objective optimization model for fenvironmental benign process planning. Proceedings of 16th International Conference on Industrial Engineering and Engineering Management, 869-873.

Kellens, K. (2013): Energy and resource efficient manufacturing - Unit Process Analysis and Optimisation. Dr.-Ing. Dissertation, Katholieke Universiteit Leuven, Leuven, Belgium.

Kellens, K., Dewulf, W., Overcash, M., Hauschild, M.Z., Duflou, J.R. (2012): Methodology for systematic analysis and improvement of manufacturing unit process life-cycle inventory - CO2PE! initiative (cooperative effort on process emissions in manufacturing). Part 1: Methodology description (UPLCI). The International Journal of Life Cycle Assessment, 17/1:69-78.

Kirsch, B., Effgen, C., Büchel, M., Aurich, J.C. (2014): Comparison of the embodied energy of a grinding wheel and an end mill. Procedia CIRP, 15:74-79.

Klocke, F., Schlosser, R., Tönissen, S. (2010): Prozesseffizienz durch Parameterwahl. wt Werkstattstechnik online, 100/5:346-349.

Knop, M. (1989): Rechnergeführte Stellgrößenwahl beim Außenrundeinstechschleifen. Dr.-Ing. Dissertation, RWTH Aachen, Aachen, Germany.

Kong, D., Choi, S., Dornfeld, D.A. (2013): Software Support for Environmentally Benign Mold Making Process and Operations. 20th CIRP International Conference on Life Cycle Engineering, Singapore, April 17-19, 279-284.

Kuhrke, B. (2011): Methode zur Energie- und Medienbedarfsbewertung spanender Werkzeugmaschinen. Dr.-Ing. Dissertation, Technische Universität Darmstadt, epubli GmbH, Berlin, Germany.

Kundrák J., Mamalis A.G., Gyani K., Markopoulos, A. (2006): Environmentally Friendly Precision Machining. Materials and Manufacturing Processes, 21/1:29-37.

Kuram, E., Ozcelik, B., Bayramoglu, M., Demirbas, E., Simsek, B.T. (2013): Optimization of cutting fluids and cutting parameters during end milling by using D-optimal design of experiments. Journal of Cleaner Production, 42:159-166.
Larek, R. (2012): Ressourceneffiziente Auslegung von fertigungstechnischen Prozessketten durch Simulation und numerische Optimierung. Dr.-Ing. Dissertation, Universität Bremen, Shaker Verlag, Aachen, Germany.
Li, G.F., Wang, L.S., Yang, L.B. (2002): Multi-parameter optimization and control of the cylindrical grinding process. Journal of Materials Processing Technology, 129/1-3:232-236.
Li, W. (2012): Energy and eco-efficiency of manufacturing processes. Ph.D Thesis, The University of New South Wales, Sydney, Australia.
Li, W., Kara, S. (2011): An empirical model for predicting energy consumption of manufacturing processes. Proceedings of the Institution of Mechanical Engineers, Part B: Journal of Engineering Manufacture, 225/9:1636-1646.
Li, W., Winter, M., Kara, S., Herrmann, C. (2012): Eco-efficiency of manufacturing processes: A grinding case. CIRP Annals - Manufacturing Technology, 61/1:59-62.
Linke, B., Duscha, M., Klocke, F., Dornfeld, D. (2011): Combination of Speed Stroke Grinding and High Speed Grinding with Regard to Sustainability. Proceedings of the 44th CIRP International Conference on Manufacturing Systems, June 1-3, Madison, USA.
Linke, B.S., Corman, G.J., Dornfeld, D.A., Tönissen, S. (2013): Sustainability indicators for discrete manufacturing processes applied to grinding technology. Journal of Manufacturing Systems, 32/4:556-563.
Linke, B.S., Das, J., Lam, M., Ly, C. (2014): Sustainability Indicators for Finishing Operations based on Process Performance and Part Quality. Procedia CIRP, 14:564-569.
Liu, Q., Chen, X., Gindy, N. (2008): Robust design and optimisation of aerospace alloy grinding by different abrasive wheels. The International Journal of Advanced Manufacturing Technology, 11-12:1125-1135.
Mori, M., Fujishima, M., Inamasu, Y., Oda, Y. (2011): A study on energy efficiency improvement for machine tools. CIRP Annals - Manufacturing Technology, 60/1:145-148.
Morrow, W.R., Qi, H., Kim, I., Mazumder, J., Skerlos, S.J. (2004): Environmental aspects of laser-based and conventional tool and die manufacturing. Journal of Cleaner Production, 15/10:932-943.
Munoz, A.A., Sheng, P. (1995): An analytical approach for determining the environmental impact of machining processes. Journal of Materials Processing Technology, 53/3–4:736-758.
Murray, V.R., Zhao, F., Sutherland, J.W. (2012): Life cycle analysis of grinding: a case study of non-cylindrical computer numerical control grinding via a unit-process life cycle inventory approach. Proceedings of the Institution of Mechanical Engineers, Part B: Journal of Engineering Manufacture, 226/10:1604-1611.
Narita, H., Kawamura, H., Norihisa, T., Chen, L., Fujimoto, H., Hasebe, T. (2006): Development of Prediction system of environmental burden for machine tool operation. JSME International Journal Series C, 49/4:1188-1195.
Narita, N., Desmira, N., Fujimoto, H. (2008): Environmental burden analysis for machining operation using LCA method. Manufacturing Systems and Technologies for the New Frontier, The 41st CIRP Conference on Manufacturing Systems, May 26-28, 65-68.
Oliveira, J.F.G., Alves, S.M. (2006): Development of Environmentally Friendly Fluid for CBN Grinding. CIRP Annals - Manufacturing Technology, 55/1:343-346.
Osterhaus, G. (1994): Verfahrensübergreifende Simulation und Auslegung von Schleifprozessen. Dr.-Ing. Dissertation, RWTH Aachen, Shaker Verlag, Aachen, Germany.
Pušavec, F., Krajnik, P., Kopač, J. (2010): Transitioning to sustainable production – Part I: application on machining technologies. Journal of Cleaner Production, 18/2:174-184.
Rajemi, M.F., Mativenga, P.T., Aramcharoen, A. (2010): Sustainable machining: selection of optimum turning conditions based on minimum energy considerations. Journal of Cleaner Production, 18/10–11:1059-1065.
Reinhardt, S. (2013): Bewertung der Ressourceneffizienz in der Fertigung. Technische Universität München, Herbert Utz Verlag, München, Germany.

Salonitis, K., Tsoukantas, G., Drakopoulos, S., Stavropoulos, P., Chryssolouris, G. (2006): Environmental Impact Assessment of Grind-Hardening Process. The 13th CIRP International Conference on Life Cycle Engineering, Katholieke Universiteit Leuven, Belgium, May 31 - June 2, 657-662.

Schiefer, E. (2001): Ökologische Bilanzierung von Bauteilen für die Entwcklung umweltgerechter Produkte am Beispiel spanender Fertigungsverfahren. Dr.-Ing. Dissertation, Technische Universität Darmstadt, Shaker Verlag, Aachen, Germany.

Schlosser, R. (2013): Methodik zur Prognose der Nachhaltigkeit des Energie- und Stoffeinsatzes spanender Fertigungsprozesse. Dr.-Ing. Dissertation, RWTH Aachen, Shaker Verlag, Aachen, Germany.

Schlosser, R., Klocke, F., Lung, D. (2011): Sustainability in Manufacturing - Energy Consumption of Cutting Processes. Advances in Sustainable Manufacturing - Proceedings of the 8th Global Conference on Sustainable Manufacturing, 85-90.

Schultz, A. (2002): Methode zur integrierten ökologischen und ökonomischen Bewertung Produktionsprozessen und -technologien. Dr.-Ing. Dissertation, Otto-von-Guericke-Universität Madgeburg, Madgeburg, Germany.

Shin, S.J. (2009): Development of a Framework for Green Productivity Enhancement. Ph.D Thesis, Pohang University of Science and Technology, Pohang, South Korea.

Soković, M., Mijanović, K. (2001): Ecological aspects of the cutting fluids and its influence on quantifiable parameters of the cutting processes. Journal of Materials Processing Technology, 109/1–2:181-189.

Srinivasan, M., Sheng, P. (1999a): Feature-based process planning for environmentally conscious machining - Part 1: microplanning. Robotics and Computer-Integrated Manufacturing, 15/3:257-270.

Srinivasan, M., Sheng, P. (1999b): Feature based process planning for environmentally conscious machining - Part 2: macroplanning. Robotics and Computer-Integrated Manufacturing, 15/3:271-281.

Thiede, S. (2012): Energy Efficiency in Manufacturing Systems. Dr.-Ing. Dissertation, Technische Universität Braunschweig, Springer Verlag, Berlin, Germany.

Wang, Q., Liu, F., Wang, X. (2014): Multi-objective optimization of machining parameters considering energy consumption. The International Journal of Advanced Manufacturing Technology, 71/5-8:1133-1142.

Wen, X.M., Tay, A.A.O., Nee, A.Y.C. (1992): Micro-computer-based optimization of the surface grinding process. Journal of Materials Processing Technology, 29/1-3:75-90.

Wolfram, F. (1986): Aspekte der energetischen Bewertung von Produkten und Prozessen der Abtrenntechnik nach den Prinzipien der vergegenständlichen Energie. Dr.-Ing. Dissertation, Technische Hochschule Karl-Marx-Stadt, Karl-Marx-Stadt, German Democratic Republic.

Xue, H., Kumar, V., Sutherland, J.W. (2007): Material flows and environmental impacts of manufacturing systems via aggregated input–output models. Journal of Cleaner Production, 15/13-14:1349-1358.

Yan, J., Feng, C., Li, L. (2014): Sustainability assessment of machining process based on extension theory and entropy weight approach. The International Journal of Advanced Manufacturing Technology, 71/5-8:1419-1431.

Yan, J., Li, L. (2013): Multi-objective optimization of milling parameters – the trade-offs between energy, production rate and cutting quality. Journal of Cleaner Production, 52:462-471.

Zein, A. (2012): Transition Towards Energy Efficient Machine Tools. Dr.-Ing. Dissertation, Technische Universität Braunschweig, Springer Verlag, Berlin, Germany.

Zhang, H., Haapala, K.R. (2014): Integrating sustainable manufacturing assessment into decision making for a production work cell. Journal of Cleaner Production, In Press.

Zhao, F., Bernstein, W.Z., Naik, G., Cheng, G.J. (2010): Environmental assessment of laser assisted manufacturing - case studies on laser shock peening and laser assisted turning. Journal of Cleaner Production, 18/13:1311-1319.

Chapter 4
Concept Development

Abstract Building on the information provided in the previous chapter, this one propose a concept which makes it possible to describe, model, evaluate and improve the grinding process and system. The concept shall support the planning of new or the improvement of existing grinding processes and systems. For this purpose, the requirements and characteristics of the proposed concept are presented in the first step. Followed by a general description of the proposed concept, each concept element is presented in detail. This detailed presentation includes the illustration of the process and system, the specification of the procedure used to model the technological, economic and environmental indicators, and the evaluation as well as the improvement procedures. The chapter ends with the presentation of the proposed application cycle to evaluate and improve the eco-efficiency of grinding processes and systems.

4.1 Concept Requirements and Characteristics

4.1.1 Concept Requirements

The review of the state of the art indicated an insufficient emphasis of the antagonistic relationships between the technological, economic and environmental indicators during the planning and improving of new or existing metal removal processes. The review especially pointed towards an absence of decision support for processes with geometrically undefined cutting edges. Accordingly, a research demand was derived and described in the previous section.

Therefore, a concept which allows the description, modelling, evaluation and improvement of the eco-efficiency of grinding processes and systems is proposed. The concept covers the following:

- the independency from a specific grinding process and system configuration, due to a generic approach,
- the influence of changing grinding process and system parameters and properties as well as grinding system configurations,

M. Winter, *Eco-efficiency of Grinding Processes and Systems*,
Sustainable Production, Life Cycle Engineering and Management,
DOI 10.1007/978-3-319-25205-6_4

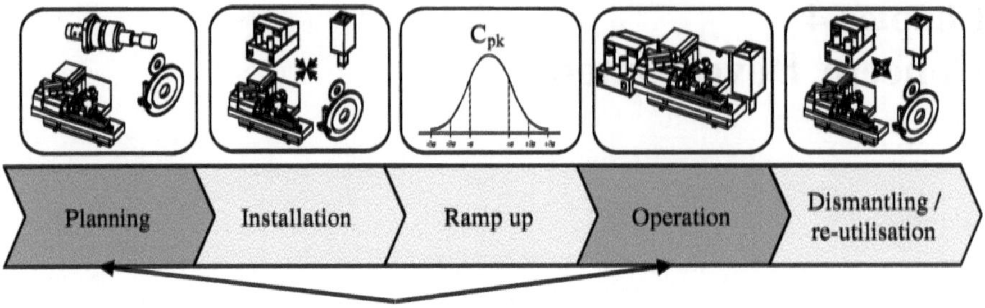

Fig. 4.1 Application focus of the proposed concept within the grinding process and system life phases (according to Schenk et al. 2014, pp. 151)

- all relevant energy and resource flows as well as their interdependencies at grinding process and system level and
- a user-demand-oriented single- as well as multi-objective evaluation and improvement of the grinding process and system.

The approach can be applied to increase the eco-efficiency of new or already-operating grinding processes and systems (Fig. 4.1). To meet this aim, a mixed deductive–inductive approach was elaborated. The reason for the elaboration of a mixed deductive–inductive approach is due to the application of physical models and empirical models. The physical model represents the deductive approach, while the empirical model represents an inductive approach. The figure shows the application focus of the proposed concept during the grinding process and system's life phases (according to Schenk et al. 2014, pp. 151). During the planning and operation phase, the proposed concept allows either the evaluation of different grinding process and system configurations or the improvement of existing configurations. In both cases, the proposed concept entails the consideration of technological, economic and environmental indicators to achieve increased eco-efficiency.

4.1.2 Concept Characteristics

To address the requirements previously illustrated, this section presents the characteristics of the proposed concept. For this purpose, Fig. 4.2 shows the relation between the criteria (left side) from Table 3.1 and the characteristics of the proposed concept (right side). One characteristic can address several criteria. The number of connections to different criteria does not indicate that the characteristic is more important as they are all equally important. The characteristics are briefly explained in the following parts.

- **Grinding types.** To cover a broad spectrum of different grinding process situations, it is important to consider different grinding types. This includes the

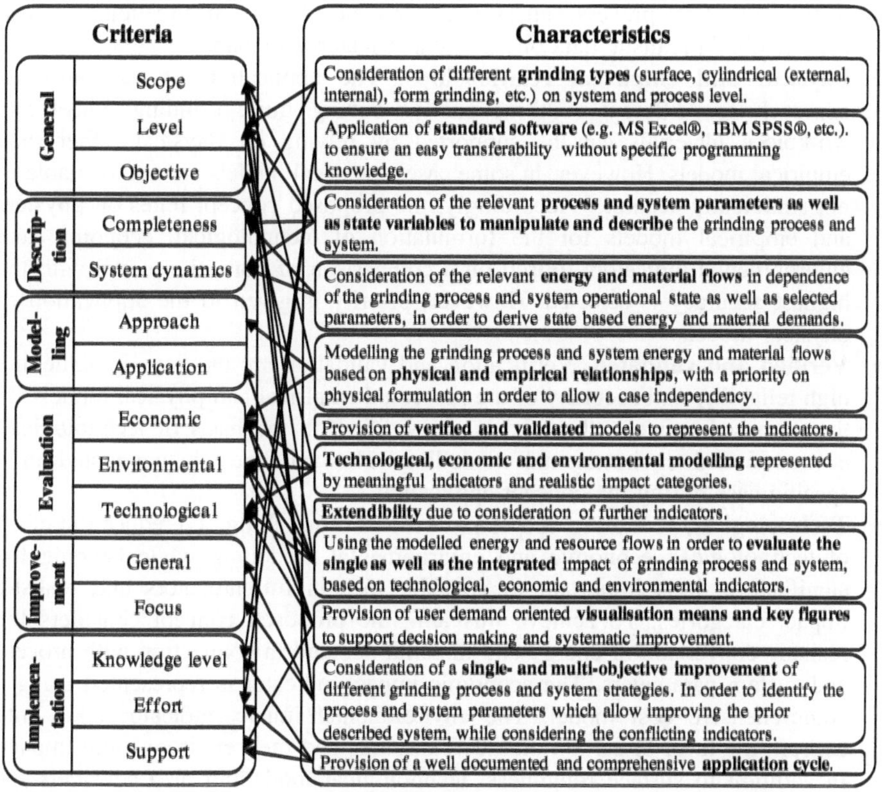

Fig. 4.2 Relationships between criteria and characteristics of the proposed concept

commonly applied surface and cylindrical grinding process types as well as centreless grinding or gear grinding. Accordingly, the proposed concept focuses on the respective process and system characteristics and features.

- **Standard software.** To support the application of the proposed approach, it is crucial to use software which is accessible and utilisable without specific programming knowledge or high expenses. To reach this aim, the proposed concept applies standard software products (e.g. MS Excel®, IBM SPSS®) to model, evaluate and improve the grinding process and system. However, user specific software can also be used to implement the proposed concept.
- **Process and system parameters as well as properties to manipulate and describe.** Section 2.1 shows that multiple process and system parameters as well as properties need to be considered in order to comprehensively describe the grinding process and system. This requirement is achieved through the application of an equation based and therewith parametrisable grinding process and system model.
- **Energy and material flows.** The operation of the grinding process and system is connected with the demand of energies and materials. The proposed concept considers all relevant energy and material flows and formulates them in

operational state; process and system parameter-dependent equations are also used to obtain realistic data on energy and material demand.

- **Physical and empirical relationships**. The application of the proposed approach for different combinations of cutting fluid, grinding wheel and workpiece materials can be supported by relying on physical rather than empirical models. However, in some cases, physical models are not available or not sufficiently accurate. Accordingly, the proposed concept relies on physical and empirical models for the formulation of technological, economic and environmental indicators. Based on a comparison of both, the model with the highest accuracy is selected. However, priority is given to the application of physical models.
- **Verified and validated**. Verification and validation are the keys in achieving high reliability and accuracy of the presented empirical and physical models. In the proposed concept, the models are verified on the basis of technological, economic and environmental modelling. The validation is conducted in a specific case during the concept application.
- **Technological, economic and environmental modelling**. To reduce the effort of technological, economic and environmental modelling while keeping the significance, it is important to consider meaningful indicators and realistic impact categories. To achieve this aim, the proposed concept considers the realistic conversion of energy and material demand in connection with process and system parameters. The economic impact indicator is represented using a comprehensive cost model. The environmental impact indicator takes into account the level of carbon dioxide emissions and their environmental impact. In addition to surface roughness, technological modelling also considers the grinding forces and the demand of power, as well as grinding wheel and cutting fluid as indicators.
- **Extendibility**. Not every user has the same technological, economic and environmental indicators. Therefore, the proposed concept provides an extendable structure which allows the inclusion of further indicators in addition to the existing ones.
- **Single as well as integrated evaluation**. Depending on the considered application scenario, the concept allows the evaluation of single as well as integrated influences of the modelled energy and material flows on the grinding process and system's technological, economic and environmental indicators.
- **Visualisation means and key figures**. To support decision-making and systematic improvement, the proposed concept provides user-demand-specific key figures and visualisation means, based on the results of the single and integrated evaluation. Furthermore, the used modelling and evaluation approach allows a detailed breakdown of the influence of the cutting fluid, electrical energy, grinding wheel and filtration material demand.
- **Single- and multi-objective improvement**. An improvement step is included within the proposed concept in order to identify the process and system parameters, while considering the conflicting indicators. This step allows the improvement of the previously described modelled and evaluated system. The

approach includes single as well as multi-objective improvements of the technological, economic and environmental indicators.

- **Application cycle**. To support the application of the proposed concept, it is important to provide a well-documented and comprehensive application cycle. Accordingly, a step-by-step methodology is provided to support the description, modelling, evaluation and improvement of the grinding process and system with the goal of achieving higher eco-efficiency. The guided methodology is implemented in a prototypical tool.

4.2 Concept Structure

Considering the previously introduced requirements and characteristics and to bridge the identified gap in the description, modelling, evaluation and improvement of eco-efficiency during the planning of new or the improvement of existing grinding processes and systems, a conceptual framework is developed and presented in Fig. 4.3. The proposed framework is composed of four elements and follows the basic idea of the Plan, Do, Check, Act (PCDA) cycle (Deming 2000,

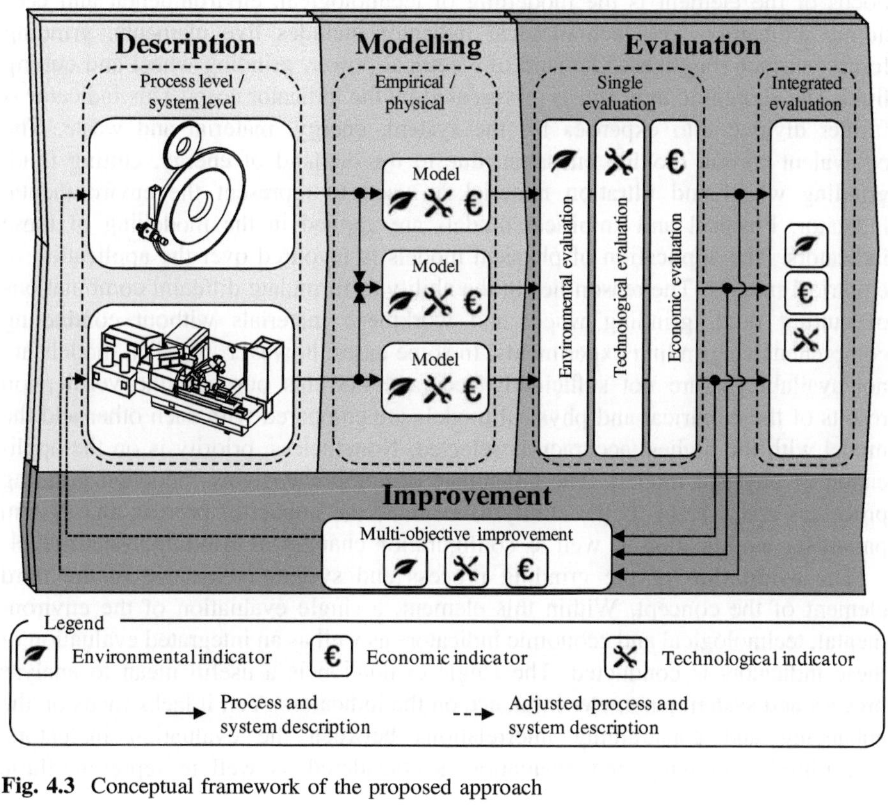

Fig. 4.3 Conceptual framework of the proposed approach

pp. 88) as well as the control loop presented in Fig. 2.11. These elements describe: the investigated grinding process and system, the empirical and physical modelling of the considered environmental, economic and technological indicators, the evaluation of the indicators with regard to their respective single as well as integrated impact and the multi-objective improvement of the grinding process and system.

The first element focuses on the **description** of the grinding process and system. For this purpose, the grinding process and system is described separately according to the necessary properties and parameters for the process and system level. The process level includes the description of the cutting fluid, the grinding wheel, the workpiece and the relevant parameters. The system level specifies the properties of the grinding machine, the cutting fluid filter, the exhaust air filter and the boundary conditions. The properties are clustered on the basis of their connection with the technological, economic and environmental indicators. The parameters are distinguished in terms of their process influence on chip formation and systems influence. As a result, the element description allows a clear classification of properties and parameters according to the contents of the two levels and creates the foundation of the successive elements.

The second element aims at the **modelling** of the grinding process and system. The modelling is based on properties and parameters from the description element. Focus of the element is the modelling of technological, environmental and economic indicators. The technological indicator includes five elements: grinding forces, surface roughness, demand of electrical power, grinding wheel and cutting fluid. The economic indicator is represented by the indicator costs. This indicator is further divided into expenses for the system, energy, material and waste. The equivalent carbon dioxide emission, due to the demand of energy, cutting fluid, grinding wheel and filtration material, is used to represent the environmental indicator. Physical and empirical models are applied in the modelling of these indicators. The application of physical models is favoured over the application of empirical models. The reason lies in the ability to formulate different combinations of cutting fluid, grinding wheel and workpiece materials without conducting comprehensive grinding experiments. In some cases, however, physical models are not available or are not sufficiently accurate. For this purpose, the verification results of the empirical and physical models are compared with each other and the model with the highest accuracy is selected. Nonetheless, priority is on the application of physical models. The advantage of comprehensively modelled grinding processes and systems is the ability to evaluate the impact of process and system parameter modification as well as configuration changes at grinding system level.

The **evaluation** of the grinding process and system is realised in the third element of the concept. Within this element, a single evaluation of the environmental, technological and economic indicators as well as an integrated evaluation of these indicators is conducted. The single evaluation is a useful mean to analyse process and system parameter influence on the indicators. Yet, it lacks focus on the enhancing and constraining interrelations between the evaluation indicators. Accordingly, an integrated evaluation is considered as well to represent these relationships. To this end, the eco-efficiency assessment approach according to DIN EN ISO 14045 is applied.

The last element focuses on the **improvement** of the grinding process and system. It focuses on the improvement of the initially described, modelled and evaluated grinding process and system through the derivation of technological and organisational measures. A multi-objective algorithm is applied to minimise the technological, economic and environmental impact based on a variation of the process and system parameter. The improvement is conducted taking into account the conflicts among the indicators.

An application cycle is included within the conceptual approach. Within nine steps, the user should be supported by the application cycle in regard to a purposeful application of the concept to bridge the identified gap.

The four elements, as well as the application cycle, are described in more detail in the following sections beginning with the description of the grinding process and system as a foundation for the subsequent empirical and physical modelling, the single and integrated-impact evaluation and the improvement element.

4.3 Process and System Description

The description of the grinding process and system's configuration can be realised using process and system parameters as well as properties. The parameters include relationships to specify the cutting motion and infeed motion to achieve continuous chip removal (Klocke 2009, pp. 251) as well as the supply of cutting fluid and the removal of exhaust air. Properties include physical material properties, concentrations and geometries. Depending on the considered hierarchical level of grinding, different parameters and properties can be considered (Fig. 2.1). Only parameters and properties crucial for the proposed technological, economic and environmental modelling are presented in Sect. 4.4. As a result of this orientation, appropriate modelling and eco-efficiency can be achieved. Further parameters and properties can be considered if additional modelling indicators are included.

Process level description

Figure 4.4 shows the describing parameters and properties for the first and second hierarchical levels in grinding, chip formation and grinding process. The figure is structured into properties of the grinding wheel and dresser, the workpiece and the cutting fluid as well as the parameters of the grinding process. The properties are further classified according to their relevance with regard to the technological, economic and environmental modelling indicators. The parameters are classified in terms of their relation to chip formation or cutting fluid and exhaust air supply or extraction.

The description of the technological properties of the **grinding wheel and conditioning unit** includes the physical material properties of the abrasive grain's bulk material hardness ($H_{b,g}$), the grain surface's material hardness ($H_{s,g}$), the

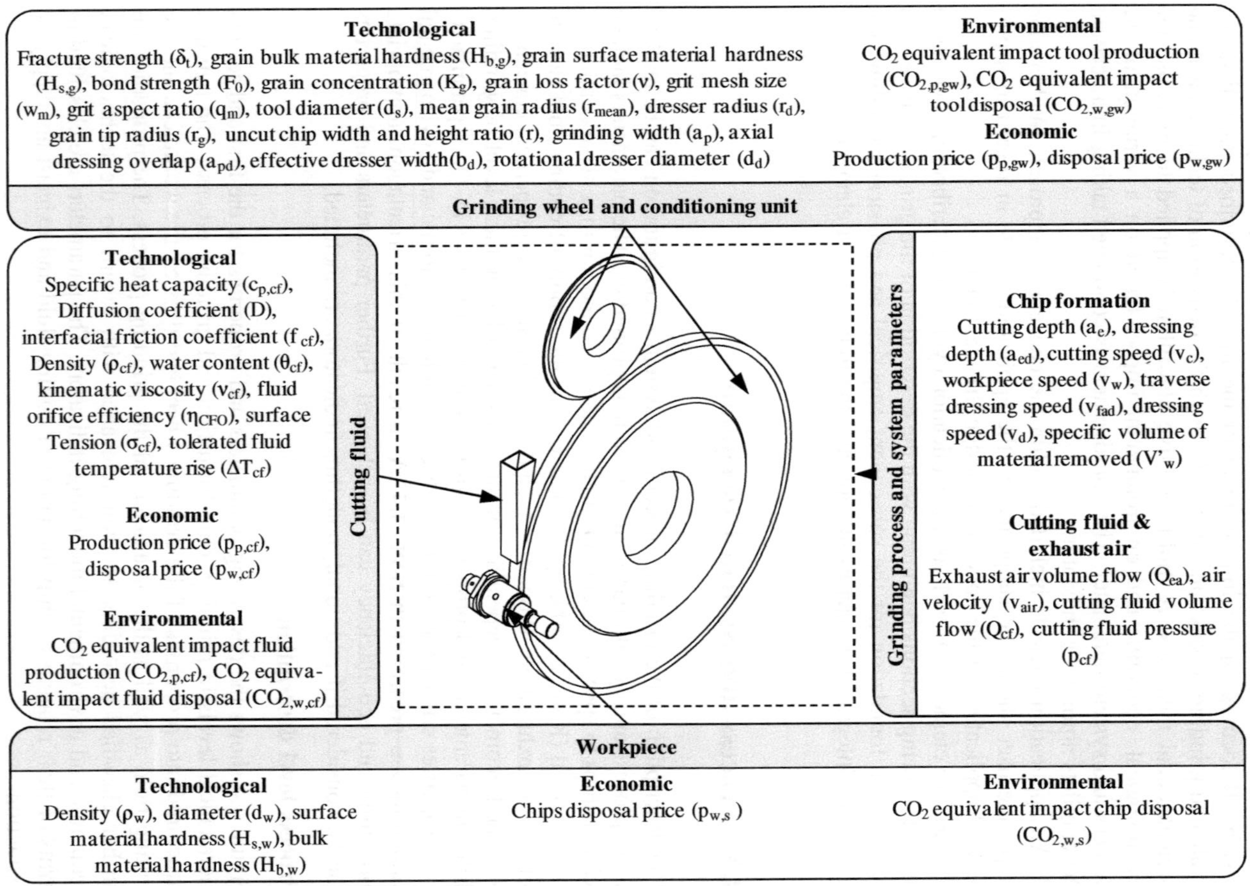

Fig. 4.4 Describing parameters and properties of chip formation and the grinding process

tensile fracture strength (δ_t) and the bond strength (F_0). In addition to the physical grain's state, grain concentration (K_g) and grain loss factor (ν) also need to be considered. The geometries of grain [average grit mesh size (w_m), grit aspect ratio (q_m), mean grain radius (r_{mean}) and grain tip radius (r_g)], grinding wheel diameter (d_s) and dresser [axial dressing overlap (a_{pd}), effective dresser width (b_d), grinding width (a_p), radius of the dresser (r_d) and rotational dresser diameter (d_d)] are also considered. Furthermore, the ratio between uncut chip width and height (r) is used. The economic and environmental properties include the impact of price and CO_2 resulting from the production ($p_{p,gw}$ and $CO_{2,p,gw}$) and disposal ($p_{w,gw}$ and $CO_{2,w,gw}$) of the grinding wheel.

The **workpiece** can be technologically described using workpiece bulk material hardness ($H_{b,w}$), workpiece surface material hardness ($H_{s,w}$), density (ρ_w) and, in case of cylindrical geometries, workpiece diameter (d_w). The price ($p_{w,s}$) and the CO_2 equivalent impact ($CO_{2,w,s}$) to dispose of the workpiece-related grinding swarf is used to describe the economic and environmental impact.

The technological properties describing the **cutting fluid** include predominantly physical (mechanical, thermal and material) fluid characteristics. The mechanic characteristics include kinematic viscosity (ν_{cf}), surface tension (σ_{cf}) and the interfacial friction coefficient (f_{cf}). Specific heat capacity ($c_{p,cf}$) and tolerated fluid temperature rise (ΔT_{cf}) are used in describing thermal fluid characteristics. The efficiency of the cutting fluid application is represented with fluid orifice efficiency (η_{CFO}). The cutting fluid material properties include density (ρ_{cf}), diffusion coefficient (D) and water content (θ_{cf}). The cutting fluid's economic and environmental properties include price and CO_2 equivalent impact resulting from the production ($p_{p,cf}$ and $CO_{2,p,cf}$) and disposal ($p_{w,cf}$ and $CO_{2,w,cf}$).

Grinding process parameters were selected on the basis of a literature review and allow a consistent description of the grinding process' relationships while reducing the modelling effort (Klocke 2009, pp. 161ff.; Knop 1989, pp. 24f.; Czenkusch 2000, pp. 38ff.). Respective parameters are the cutting speed (v_c) of the grinding wheel and the dressing speed (v_d) if a rotational conditioning unit is used. Both parameters allow the influence of equivalent chip thickness and connected tool wear, grinding forces as well as grinding wheel and workpiece roughness. The traverse dressing speed (v_{fad}) and dressing depth (a_{ed}) have a further impact on the achievable roughness. Workpiece speed (v_w) and cutting depth (a_e) determine the material removal rate of the grinding process and thus the overall productivity of the process. When working with varying quantities of ground workpiece material, the specific volume of material removed (V'_w) needs to be taken into account. To model the interaction between grinding process and system, the input and output flows of the three sub-systems (grinding machine, cutting fluid filter and exhaust air filter) can be used. The cutting fluid's volume flow (Q_{cf}) and needed pressure (p_{cf}) allows a link between cutting fluid filter and power demand for the grinding process and system, as well as cutting fluid demand. Exhaust air volume flow (Q_{ea}) and air velocity (v_{air}) can be used to predict the cutting fluid demand.

System level description

As shown in Fig. 2.1, the next hierarchical level is the grinding system. The necessary properties to model the technological, economic and environmental indicators of the grinding system are presented in Fig. 4.5. The figure classifies the properties of the grinding machine, the cutting fluid filter, the exhaust air filter and the boundary conditions according to their relevance to the technological, economic and environmental modelling indicators.

The **grinding machine's** technological properties include variables to describe the power and cutting fluid-related demand behaviour. The power behaviour is determined using the electrical power demand ($P_{GM,i}$) of the grinding machine components and the compressed air volume flow (Q_{ca}). Cutting fluid behaviour is characterised using the machine tool's interior temperature (T_{GM}) and the surface area of evaporating cutting fluid (A_{cf}). Economic properties include grinding machine-hour rate ($p_{m,GM}$) time for tool travel (t_t) and workpiece handling (t_h).

The technological properties of the **cutting fluid filter** can be classified into variables describing power and cutting fluid-related demand behaviour. The power demand depends on the cutting fluid filter's power demand ($P_{CFF,i}$) and the electric pump's drive efficiency (η_{el}). The behaviour of cutting fluid demand varies according to the surface areas of the used filtration material (A_{fm}) as well as on the mean cutting fluid temperature ($\overline{T}_{cf}$). The prices for filtration material procurement ($p_{p,fm}$) and disposal ($p_{w,fm}$) and the machine-hour rate ($p_{m,CFF}$) can be used as economic properties. Connected with those variables is the environmental property's equivalent CO_2 impact resulting from filtration material production ($CO_{2,p,fm}$) and disposal ($CO_{2,w,fm}$).

The properties of the **exhaust air filter** include the exhaust air filter's power demand ($P_{EAF,i}$) and the machine-hour rate ($p_{m,EAF}$). The **boundary conditions** include the technological properties for ambience air temperature (T_a) and humidity (ϕ_{air}) as well as the specific compressor power demand for compressed air generation (P'_{ca}). Connected with the electrical energy demand is the energy price (p_e) and equivalent CO_2 impact due to electrical energy production ($CO_{2,p,e}$). The price of labour (p_l) is considered as a further economic state variable.

These parameters and properties allow the modelling of technological, economic and environmental indicators. Corresponding values of these parameters and properties can be obtained from the investigated grinding system, from specific literature as well as from the next section and the appendix.

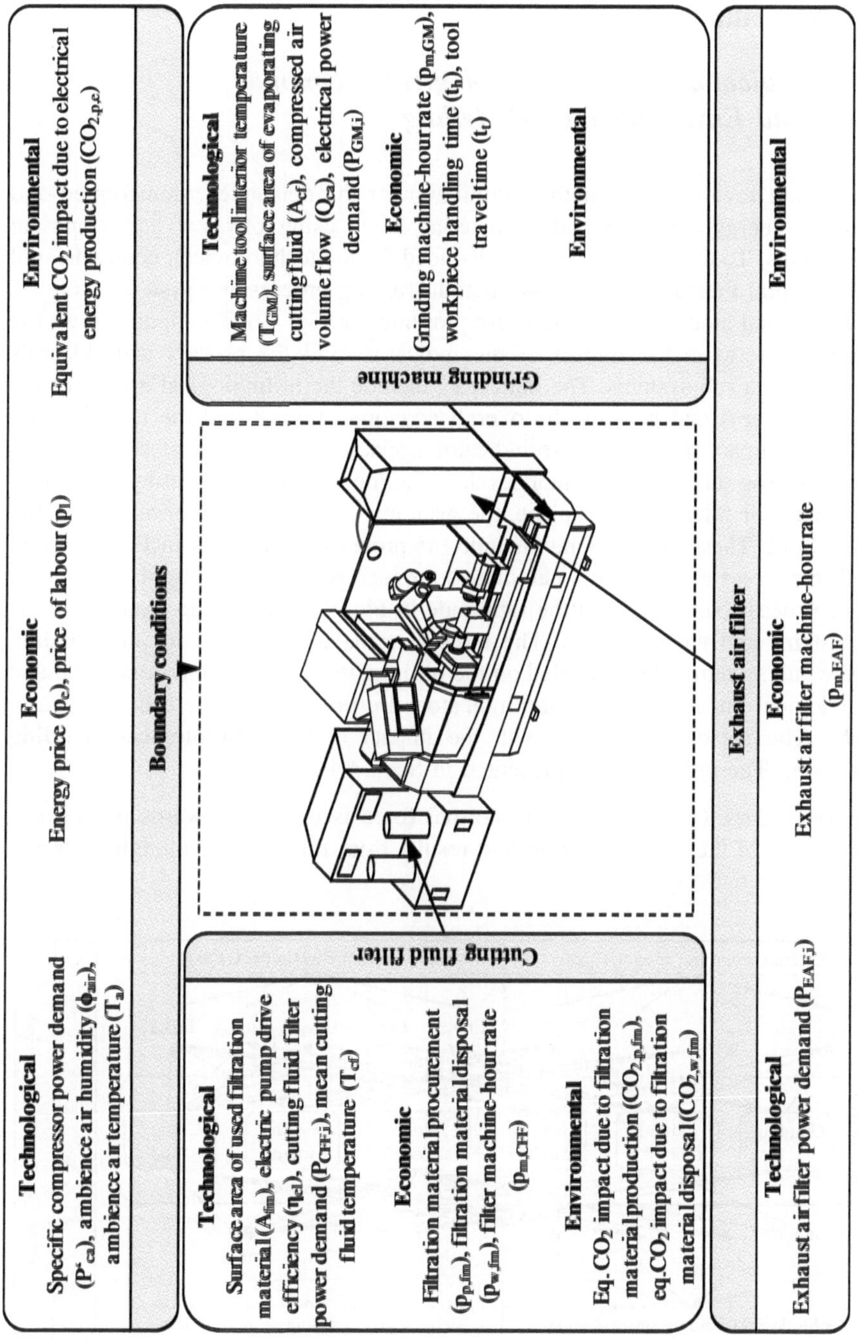

Fig. 4.5 Describing properties of the grinding system

4.4 Technological, Economic and Environmental Modelling

4.4.1 Indicators of the Technological, Economic and Environmental Modelling

This section revolves around the formulation of the demanded, transformed and generated energy, material and immaterial flows using physical and empirical relationships. To this end, models are deduced for the technological, economic and environmental indicators of the considered grinding process and system level.

The derived modelling indicators are presented in Fig. 4.6. The figure shows the connection between the impact of the indicators and the process and system's parameters via sub-systems. The indicators include the technological impact of the grinding process and system, the overall economic impact and the overall environmental impact. Each of these indicator impacts is composed of several other indicators. The sub-systems comprises the grinding machine, the cutting fluid filter, the exhaust air filter, the workpiece, the grinding wheel and dresser as well as the cutting fluid. These six elements include the properties presented in Figs. 4.4 and 4.5. The process and system's parameters are also presented in Fig. 4.4.

The process parameters on the right side in Fig. 4.6 are used to manipulate the sub-systems and the indicators. The indicators make it possible for the grinding process and system to be modelled on the basis of its economic, environmental and technological impact. The selected indicators are presented in the following paragraphs. The presentation starts with the impact of the **technological grinding process** F_t. The indicators are presented in Sect. 4.4.2:

- *normal force* (F_n). The specific normal force is oriented towards the vertical direction of the chip formation and results from rubbing and ploughing forces

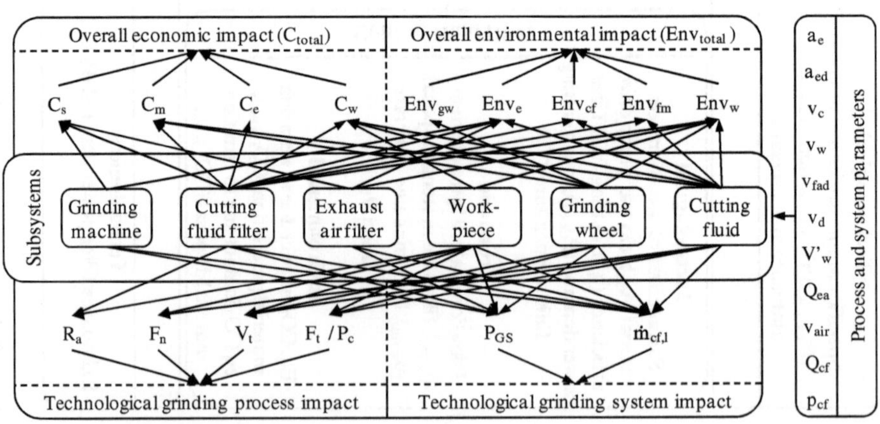

Fig. 4.6 Derived modelling indicators

during the chip formation process, as well as the hydrostatic and hydrodynamic pressure of the cutting fluid. F_n has influence on the grinding process and workpiece dimensional accuracy (Sect. 4.4.2.2).

- *cutting power* (P_c)/*the tangential force* (F_t). P_c is related by the cutting speed to F_t. F_t is oriented towards the horizontal direction of the chip formation and results from the actual cutting process. The cutting power is an important indicator of the energy/heat input into the workpiece surface and allows the prediction of workpiece temperature and changes in surface integrity (Sect. 4.4.2.2).
- *arithmetical mean roughness* (R_a). The workpiece surface's roughness is created as a result of a superposition of abrasive grain engagement tracks on the workpiece surface. The roughness, as an element of surface integrity, is an important criterion in the assessment of the quality of the workpiece (Sect. 4.4.2.3).
- *volume of tool wear* (V_t). The specific volume of tool wear describes the overall loss of abrasive material during the grinding process. V_t determines the grinding wheel's service life, the workpiece surface's quality, and process stability as well as dressing strategy and time (Sect. 4.4.2.4).

The **technological grinding system's impact** is characterised with regard to its energy and material demand. The following indicators are formulated:

- *grinding system power demand* (P_{GS}). It includes the electrical power demand of the grinding machine and the filtration systems. Accordingly, P_{GS} has a major influence of the environmental and economic impact of the grinding process and system as well (Sect. 4.4.2.5).
- *cutting fluid demand due to loss* ($\dot{m}_{cf,l}$). Fluid demand is based on the drag-out via the transforming elements of the grinding process and system. The loss influences the cutting fluids' composition and process stability as well as the overall environmental and economic impact (Sect. 4.4.2.6).

The **overall economic impact** (C_{total}) is the sum of the following indicators and is described in Sect. 4.4.3:

- *system cost* (C_s). Costs for labour, machine-hour rate of grinding machine and filtration units as well as overhead cost (e.g. transportation).
- *material cost* (C_m). Procurement costs for cutting fluid, filtration material, grinding wheel as well as overhead costs (e.g. detergents, cleaning cloth).
- *energy cost* (C_e). Expenses for electrical, mechanical and thermal energies as well as overhead cost (e.g. heating, light).
- *waste cost* (C_w). Costs for the disposal of cutting fluid, filtration material, grinding wheel residues and grinding swarf as well as overhead costs (e.g. used detergents, used cleaning cloths).

The **overall environmental impact** (Env_{total}) of the grinding process and system is the result of the sum of the following indicators (see Sect. 4.4.4):

- *equivalent CO_2 impact of the grinding wheel* (Env_{gw}). The application of the grinding wheel and the continuous material loss due to dressing and wear is connected with the environmental impact of the grinding wheel's production and disposal. Depending on the process parameters and the wheel type, a major impact can arise.
- *equivalent CO_2 impact of electrical energy* (Env_e). Env_e links the environmental impact of the electrical power demand with time.
- *equivalent CO_2 impact of cutting fluid* (Env_{cf}). This indicator considers the environmental impact connected to the initial filling and refilling of lost cutting fluid.
- *equivalent CO_2 impact of filtration material* (Env_{fm}). The demand of filtration material and its related environmental impact is the focus of this indicator
- *equivalent CO_2 impact of waste* (Env_w). Env_W includes environmental impact resulting from the disposal of cutting fluid, grinding wheels, grinding swarf and filtration material at the end of their respective service lives.

4.4.2 Technological Modelling of the Grinding Process

The procedure presented in this section allows the derivation of prediction models for the technological indicators (Fig. 4.7).

The foundations of technological modelling are presented in the first step. This step includes existing approaches for deriving the theoretical basis of the model, approaches for the design of experiments and for evaluating the predictive accuracy of the model as well as practical implementation. Based on these foundations, the physical and empirical force, surface roughness tool wear, cutting fluid demand and power demand modelling can be achieved. In the last step, the derived physical and

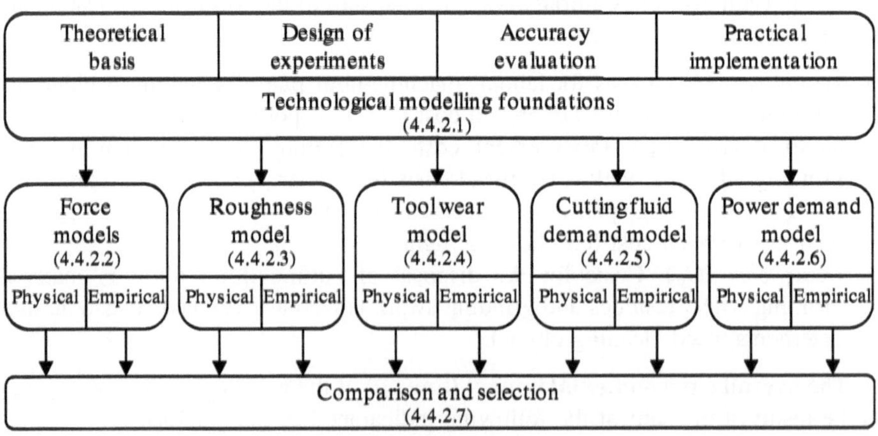

Fig. 4.7 Procedure for technological modelling (subsection numbers are in brackets)

empirical models are compared in order to select a suitable model for the prediction of technological indicators.

4.4.2.1 Foundations of Empirical and Physical Grinding Process Modelling

Theoretical basis of the proposed model

Based on the approaches used in the modelling of technological indicators presented in Sect. 2.2.3, a new model was proposed for the physical modelling of the forces, energies, workpiece surface roughness and tool wear, as well as demand of power and cutting fluid. Figure 4.8 shows the existing models and influencing factors, which were taken into account in the elaboration of the new approach.

Influencing factors are those values whose consideration allows an improved modelling accuracy for the specific force, roughness and tool wear model. Determining the number of grains is necessary to obtain the influence of the

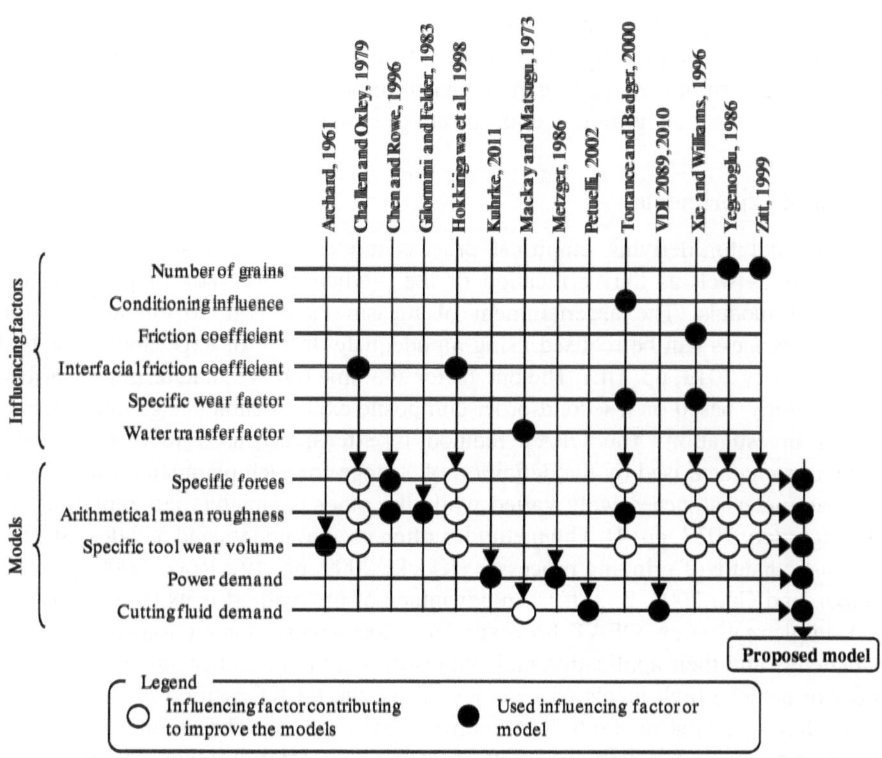

Fig. 4.8 Existing approaches taken into account for the elaboration of the proposed model

grinding wheel (e.g. grain size, concentration); for this purpose, the approaches of Yegenoglu and Zitt are used. The tool conditioning process is a further influence on the grain number. The approach proposed by Torrance and Badger thus is considered. The influence of the cutting fluid is determined using the friction coefficient (based on an approach by Xie and Williams) as well as the interfacial friction coefficient (based on the works of Challen and Oxley as well as Hokkirigawa et al.). The specific wear factor according to approaches by Xie and Williams, as well as Torrance and Badger, was included as a further influencing factor.

For the modelling of the specific grinding forces, the approach proposed by Chen and Rowe is selected in connection with elements of approaches from Challen and Oxley, Xie and Williams as well as Torrance and Badger. The arithmetical mean roughness of the workpiece surface is modelled on the basis of the approach presented by Torrance and Badger, which was initially proposed by Gilormini and Felder. Furthermore, elements of Chen and Rowe as well as Xie and Williams are included. The specific tool wear volume is based on the approach presented by Archard in combination with specific wear factors presented by Xie and Williams. The prediction of the grinding system's power demand is based on works by Kuhrke (grinding machine) and Metzger (cutting fluid filter). To model cutting fluid demand, approaches presented by Petuelli and VDI 2089 are combined and extended.

In order to increase the modelling accuracy of cutting fluid demand, the water transfer factor presented by Mackay and Matsugu was considered. The factors allow the description of cutting fluid evaporation via the exhaust filter (e.g. air velocity) as well as grinding and filtration systems (e.g. area of evaporating surface).

Design of experiments

Fundamental for deriving empirical process models is the availability of measurements as well as the verification of the predictive accuracy of physical and empirical models. The ascertainment of measurement data to characterise the grinding process can be realised using an adequate design of experiments method (Montgomery 2013, pp. 1ff.). The one factor at a time (OFAT) and response surface methodology, based on a second-order composite experimental design, are selected in this investigation. The OFAT method is extensively used in practice. This method is characterised by the definition of a range for each parameter, after which one parameter is successively varied while the other parameters are kept constant (Montgomery 2013, pp. 4). The method is often used to understand and describe the stochastic nature of grinding processes (Klocke 2009, pp. 1ff.; Rowe 2009, pp. 1ff.; Malkin and Guo 2008, pp. 1ff.). An advantage of the method is its straightforward and simple realisation. OFAT, however, lacks focus on the interactions between the parameters and their application and can result in a high number of experiments in order to achieve high levels of accuracy. Thus, the OFAT method is less efficient than other statistical methods (Montgomery 2013, pp. 5). The second-order composite experimental design is a method within the factorial experimental design. The method can be used to gain knowledge of parameter interaction while providing reasonable robustness against outliers. The method does not require a large number of experiments and represents a very efficient design (Montgomery 2013, pp. 500f.). The method was also successfully applied in the investigation of grinding processes (Li 2011, pp. 83; Winter et al. 2014, pp. 644ff.).

Evaluation of predictive accuracy

To assess the predictive accuracy of the aforementioned model, statistical methods can be applied to control how well the measured and calculated values fit. In case of regression models, the coefficient of determination (R^2) can be used. The R^2 value can range between 0 and 1. A value close to 1 indicates a high accuracy of modelled ($\widehat{Y}_i$) and observed measurement data (Y_i). The coefficient is determined using an equation (Eq. 4.1), based on modelled values ($\widehat{Y}_i$), observed values (Y_i) and mean observed values ($\overline{Y}_i$) (Fahrmeir et al. 2007, pp. 160):

$$R^2 = \frac{\sum_{i=1}^{n}\left(Y_i - \widehat{Y}_i\right)^2}{\sum_{i=1}^{n}\left(Y_i - \overline{Y}_i\right)^2} \tag{4.1}$$

A further method used to determine the statistical deviation is the relative absolute error (E). The E value is also an assessment method for the deviation between predicted and observed values. The value ranges from 0 to infinity; values close to 0 indicate a high fit. The formulation of E is based on modelled values ($\widehat{Y}_i$), observed values (Y_i) and mean observed values ($\overline{Y}_i$) (see Eq. 4.2) (Armstrong and Collopy 1992, pp. 78).

$$E = \frac{\sum_{i=1}^{n}\left|Y_i - \widehat{Y}_i\right|}{\sum_{i=1}^{n}\left|Y_i - \overline{Y}_i\right|} \tag{4.2}$$

Practical implementation

The practical implementation entails the verification of the physical and empirical models. The verification "is concerned with building the model correctly" (Banks et al. 2004, pp. 310). For this purpose, a three-step-based procedure was applied. The procedure includes the following steps: design of experiments, performing as well as analysing of the experiments and modelling. The following paragraphs introduce the used methods and materials.

Software and procedure

The design of the experiments was conducted using the MiniTab 16® software, depending on whether a factorial experimental design was chosen. The grinding processes determining cutting forces, arithmetic mean surface roughness and specific tool wear volume were repeated two times. The presented experimental results represent the mean value of these repetitions. The realisation and analysis of the experiments were performed using the LabView® software (National Instruments) to record the measurement data and the FlexPro® software (Weisang) to analyse the data. The realisation of the empirical and physical modelling was conducted using MS Excel® and IBM SPSS®. While MS Excel® was used to calculate the respective modelling indicators, IBM SPSS® was used to derive empirical models based on observed values.

Grinding process and system

The grinding experiments were conducted using a Studer S120 internal cylindrical grinding machine. The spindle has a maximum power of 10 kW and a maximum rotation speed of up to 60,000 rpm. The grinding system includes a cutting fluid filter with a filling volume of 100 l, a fluid supply pressure of 6.7 bar and a filtration volume flow of 30 l/min. Furthermore, an exhaust air filter with an exhaust air volume flow of 2500 m^3/h was connected to the grinding machine.

Grinding wheel

A vitrified bonded cubic boron nitride (cBN) grinding wheel with a medium porosity and high bond hardness was used. The wheel featured a cylindrical shape with a straight profile, the external tool diameter was 40 mm and the tool width was 15 mm (specification B126 T 43 V385). When modelling arithmetic mean surface roughness, three additional grinding wheels were used; they only differed in terms of grain size (B46, B91 and B181). The conditioning was realised with a CNC-controlled diamond form roll.

Cutting fluid

Three different cutting fluids were used during the modelling of the technological indicators. The first fluid was a mineral oil-based grinding oil. The grinding oil was a non-water miscible cutting fluid with a comparatively high viscosity of $\nu_{40°C}$ = 12.0 mm^2/s. The fluid was recommended for cutting heat-treated and case-hardened steels. The second fluid was a water miscible polymer dilution, which was recommended for high-performance grinding processes ($\nu_{40°C}$ = 4.8 mm^2/s). Moreover, the third cutting fluid was a mineral oil-based emulsion with a concentration of 7.5 % ($\nu_{40°C}$ = 1.0 mm^2/s). The fluid was especially designed for difficult steel machining processes. The cutting fluid was applied via a tangential nozzle.

Workpiece

The workpiece material consisted of hardened carbon alloy steel with the designation 1.3505 (DIN 100Cr6). The material had a hardness of 756 HV 5 and featured a ring shape. The grinding process increased the inner diameter of the ring from an initial 50 mm to a maximum diameter of 110 mm.

Measurement devices

The cutting power was metred by a three-phase power analyser (Load Controls® PPC-3) with a temporal resolution of 15 ms. The grinding tangential and normal forces were measured with two 3-component force sensors (Kistler®, type 9251A), each of which was connected to a microprocessor-controlled piezo-resistive amplifier (Kistler®, type 5747A). The signals from the power metre and the piezo-resistive amplifiers were recorded via a National Instruments® based data acquisition and processing system. The arithmetic mean surface roughness was measured at four different points on the workpiece using a surface measurement device (Hommel-Etamic® T1000 basic). The radial grinding wheel wear Δr_s was measured using a steel plate (0.25 mm) in which the grinding wheel ground after each experiment. The created grinding wheel profile was measured with a measuring probe.

4.4.2.2 Specific Normal and Tangential Forces

Characteristics of the real world

Force development depends on the chip formation process, which is a result of the geometrical as well as kinematical relationships of the grinding process. The relationships, also presented in Figs. 2.3 and 2.6, can be described using the grinding and conditioning process parameters. The effect of varying cutting speed (v_c), workpiece speed (v_W), cutting depth (a_e), traverse dressing speed (v_{fad}), specific material removed (V'_W) and dressing speed (v_d) on the specific normal (F'_n) and tangential forces (F'_t) is presented in Fig. 4.9 for an internal cylindrical grinding

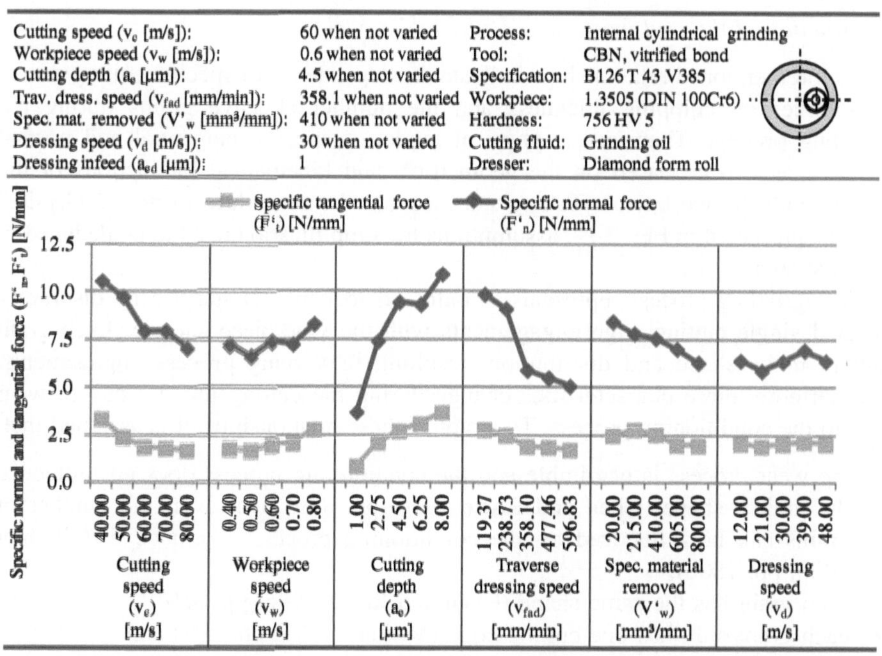

Fig. 4.9 Relationship between process parameters and specific forces in internal cylindrical grinding

process. To investigate the influence of the parameters, the OFAT method was chosen. The method was chosen to determine the influence of the changed parameters without any interference from superposing effects.

With increasing cutting speed, a decrease of the specific normal and tangential force occurs, due to the reduction of the chipping thickness. The number of engaging grains results in a reduction of the cutting edge's load. However, with the increasing cutting speed, the number of grains passing over the workpiece surface increases as well and results in higher friction (Kassen 1969, pp. 111). The higher the workpiece speed leads, if the cutting depth is constant, to a rise of the specific tangential and normal forces, due to an increase in the material removal rate. The rise of the cutting depth also results in an increase in the material removal rate, based on a growing chipping thickness and number of grains. Accordingly, the forces increase as well. As previously mentioned in Sect. 2.1.1, an increase of the traverse dressing speed leads to a rougher grinding wheel surface, resulting in a decrease of grain amount and in a decline of specific tangential and normal forces (Fig. 4.9). The increasing of the specific material removed leads to an increase in cutting time and tool wear. Higher wear reduces the number of abrasive grains and changes their shape, resulting in a reduction of forces.

To summarise, the cutting depth and the traverse dressing speed in particular have a major influence, followed by cutting and workpiece speed as well as specific material removed and dressing speed. Therefore, a model to determine the forces needs to consider these impacts.

Abstraction of the real world into a physical model

The aforementioned relationships indicate the dependence of specific cutting power and forces on chipping thickness and the number abrasive grains during the grinding process. Therefore, a physical model of specific tangential and normal forces needs to be based on the geometrical and kinematical interactions of the grains with the workpiece. Taking into account the procedure to derive physical models presented in Fig. 2.12, assumptions for simplification need to be deduced in the first step.

The grinding process represents an enforced sequence of statistically disorderly located single cutting edge engagements with the workpiece surface. Each grain differs in its shape and distribution, resulting in varying process engagements. Furthermore, these characteristics change during the cutting process due to wear and to the conditioning process. To simplify these relationships, it is assumed that:

- the wear process is negligible and the conditioning process does not influence the grain's shape (Chen and Rowe 1996, pp. 889). However, the number of grains can be influenced by the conditioning process (Torrance and Badger 2000, pp. 1806),
- each grain has the same shape (Chen and Rowe 1996, pp. 889),
- each grain only has one cutting edge (Verkerk 1977, pp. 393) and
- no plastically deformation processes occur and the entire material is removed during the cutting process (Zitt 1999, pp. 137).

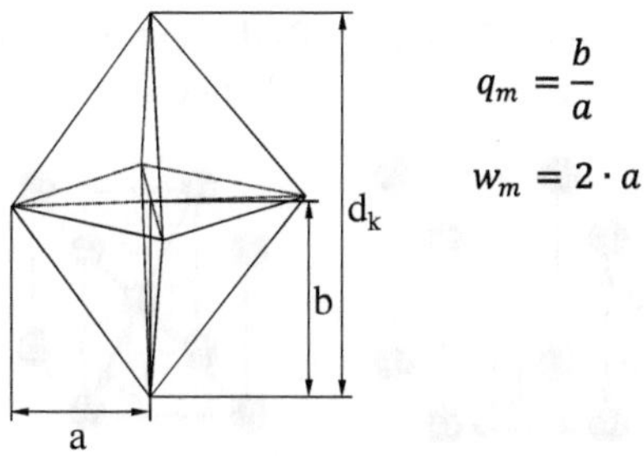

Fig. 4.10 Geometrical relationships of an octahedrally shaped grain model

Based on these simplifications, in the next step basic equations can be derived to describe the abrasive grain's shape, grain density per unit area, the specific forces and friction coefficients.

Abrasive grains commonly have a block-like shape with straight edges and plane surfaces. Different shapes can be used for the geometrical grain's description, including cuboid, octahedron, ellipsoid, sphere, exponential figure or double cone (Büttner 1968, pp. 27; Triemel 1976, pp. 37; Spiegel 1994, pp. 31; Zitt 1999, pp. 70). Investigations by Schleich, Yegenoglu and Zitt, however, highlighted that the octahedrally shaped grain is suitable in modelling grain distribution and grain density (Schleich 1982, pp. 29; Yegenoglu 1986, pp. 36; Zitt 1999, pp. 72). Therefore, this grain geometry is used in the following models.

Figure 4.10 presents an octahedrally shaped grain. The grain can be described by the mean grain diameter (d_k), the half grain width (a) and the half grain height (b). The mean grain diameter corresponds to the average grit mesh size (w_m). By relating grain height and grain width, the grit aspect ratio (q_m) can be calculated. For B126 grains (made of cBN), for example, the grit aspect ratio was determined statistically by Triemel and Heuer with a value of $q_m = 1.41$ and $q_m = 1.43$ (Triemel 1976, pp. 37; Heuer 1992, pp. 30).

Grain volume (V_K) can be determined (see Eq. 4.3) based on geometrical relationships.

$$V_K = \frac{q_m \cdot w_m^3}{3} \tag{4.3}$$

By relating grain concentration (K_g) within the abrasive layer of the grinding wheel with the calculated grain volume, volume-related grain density c_k can be calculated (Yegenoglu 1986, pp. 38):

$$c_k = \frac{K_g}{V_K} \tag{4.4}$$

Volume-related grain density represents the amount of grains in a specific volume of the abrasive layer. However, to determine the amount of grains on the

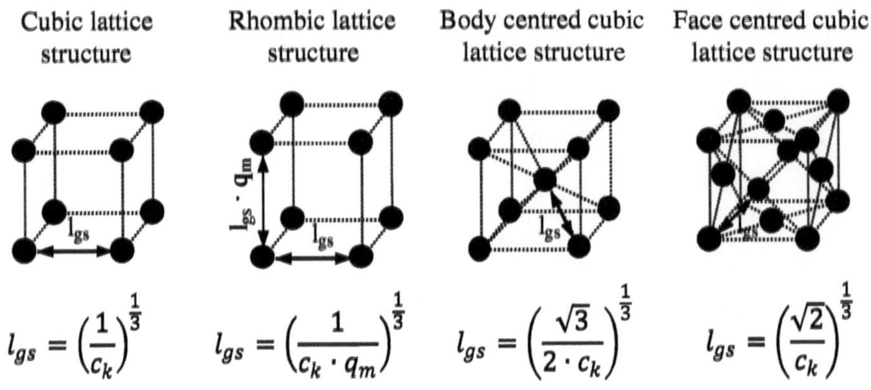

Fig. 4.11 Approaches used to describe grain spacing (according to Zitt 1999, pp. 95)

grinding wheel's surface area, the grain spacing l_{gs} has to be taken into account, to conclude from the volume to the area. The calculation of the grain's spacing in a specific volume of the abrasive layer can be performed assuming either a cubic lattice structure (Zitt 1999, pp. 93), a body-centred cubic lattice structure (Linke 1992, pp. 52; Paulmann 1994, pp. 22), a rhombic lattice structure (Yegenoglu 1986, pp. 42) or a face-centred cubic lattice structure (Schleich 1982, pp. 29) (Fig. 4.11).

In this investigation, the cubic lattice structure is considered on the basis of the previous works by Zitt (1999, pp. 93) (see Eq. 4.5).

$$l_{gs} = \left(\frac{1}{c_k}\right)^{\frac{1}{3}} \tag{4.5}$$

Therefore, the surface area-related grain density $c_k{}^*$ is calculated as follows:

$$c_k{}^* = \frac{c_k}{l_{gs}} = c_k^{\frac{2}{3}} \tag{4.6}$$

However, the determined surface area-related grain density represents a theoretical number of abrasive grains. The actual number of grains is lower and depends on the height of the grain above the bond and the number of grain breakouts (Rabiey 2010, pp. 67; Yegenoglu 1986, pp. 45). Yegenoglu proposed a grain loss factor (ν). The factor describes the actual number of grains in dependence with grinding process characteristics. On the basis of these measurements, Yegenoglu determined the grain loss factor with $\nu = 0.5$ (Yegenoglu 1986, pp. 45f.). The static number of grains N_g can be calculated as:

$$N_g = \nu \cdot c_k{}^* \tag{4.7}$$

Grain type:	CBN	Bond type:	Vitreous
Grain size:	B181	Concentration:	C200

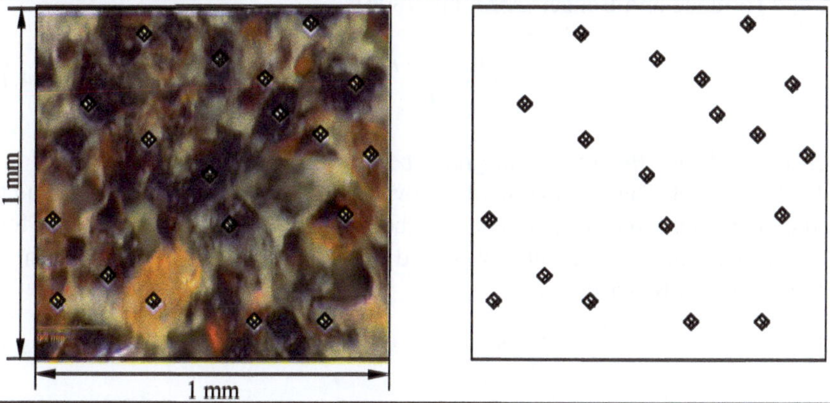

Number of grains on a surface area of 1 mm²:	Measured: 19	Calculated: 20

Fig. 4.12 Comparison of measured and calculated static number of grains

To validate the equation, Fig. 4.12 shows an example for a vitrified bonded CBN grinding wheel with grain size B181 and concentration C200. This specification results into a w_m of 165 µm, a K_g of 50 %, a ν of 0.5 and a q_m of 1.28 (based on Zitt 1999, pp. 72).

The number of grains was determined by counting them. Four different regions of the grinding wheel have been photographed using a digital optical microscope. The measured and the calculated number of grains per unit area corresponded well. Similar results were reported by Rabiey (2010, pp. 68).

The conditioning process of the grinding wheel has further influence on the number of grains per unit area. During the conditioning process, the grains are partially in contact with the dresser, resulting in a particular grain density created by the conditioning process $N_{g,d}$. Based on the investigation of Torrance and Badger, the $N_{g,d}$ can be determined as follows:

$$N_{g,d} = N_g \cdot f_s \tag{4.8}$$

f_s represents the fraction of grits contacted during one pass of the dresser and is formulated based on the dressing depth a_{ed} as well as a proportionality constant C and can be written as follows (Torrance and Badger 2000, pp. 1807):

$$f_s = \frac{1}{2 \cdot C^2 \cdot a_{ed}} \cdot \left[1 - \left(\frac{1}{exp(2 \cdot C^2 \cdot a_{ed})}\right)\right] \tag{4.9}$$

The proportionality constant C is determined based on grain fracture strength δ_t, mean grain radius r_{mean}, dresser radius r_d, bond strength F_0 and traverse dressing feed f_{ad} (Torrance and Badger 2000, pp. 1806):

$$C = \frac{42 \cdot \delta_t \cdot r_{\text{mean}} \cdot f_{ad}}{F_0 \cdot \sqrt{r_d}} \tag{4.10}$$

Once established the number of grains per unit area, the specific grinding forces can be determined. The calculation of the overall specific normal force (F'_n) and the specific tangential force (F'_t) depends on the normal ($f_{n,i}$) and tangential ($f_{t,i}$) force per grain, the number of grains ($N_{g,d}$) and the contact length between grain and workpiece. It can be written as:

$$F'_n = f_{n,i} \cdot N_{g,d} \cdot l_g \tag{4.11}$$

$$F'_t = f_{t,i} \cdot N_{g,d} \cdot l_g \tag{4.12}$$

Furthermore, both forces can be related to each other by using the friction coefficient:

$$F'_t = F'_n \cdot \mu \tag{4.13}$$

The contact length can be determined by relating the cutting depth (a_e) and the equivalent diameter (d_{eq}) (see Eqs. 4.14 and 4.15).

$$l_g = \sqrt{a_e \cdot d_{eq}} \tag{4.14}$$

$$\begin{array}{lll} d_{eq} = \frac{d_w \cdot d_s}{d_w \pm d_s} & \begin{array}{l} '+' \text{ for external cylindrical grinding} \\ '-' \text{ for internal cylindrical grinding} \end{array} \\ d_{eq} = d_s & \text{for surface grinding} \end{array} \tag{4.15}$$

During the chip formation process, the octahedrally shaped grain moves along a path determined by the diameter of the grinding wheel into the workpiece with a flat attack angle α (Fig. 4.13). After the elastic deformation of the workpiece material, an increasing plastic deformation of the workpiece material follows with increasing grain movement, resulting in an upward squeezing of the workpiece material and finally chip formation (Chen and Rowe 1996, pp. 884).

Based on the assumption that the magnitude of the force required to indent a workpiece surface is constant and independent from the direction in which it is loaded, Chen and Rowe formulated the tangential force and the normal force on a grain as (Chen and Rowe 1996, pp. 891):

$$f_{t,i} = \frac{3\pi}{4} \cdot \frac{b_{cu,max}}{h_{cu,max}} \cdot H_{b,w} \cdot \left(\frac{C'}{3}\right) \cdot (\sin\alpha - \mu\cos\alpha) \cdot A_{cu,max} \tag{4.16}$$

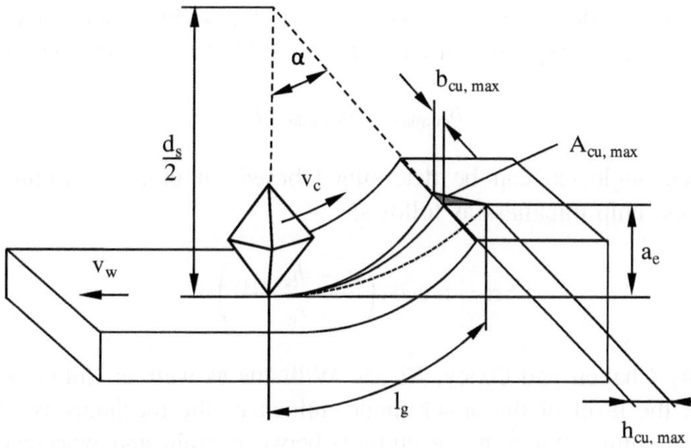

Fig. 4.13 Geometrical and kinematic relations during chip formation

$$f_{n,i} = \frac{3\pi}{4} \cdot \frac{b_{cu,max}}{h_{cu,max}} \cdot H_{b,w} \cdot \left(\frac{C'}{3}\right) \cdot (\cos\alpha - \mu \sin\alpha) \cdot A_{cu,max} \tag{4.17}$$

Both equations are based on the hardness of the bulk material ($H_{b,w}$), the friction coefficient (μ) and a constraint factor ($C' = 3$) as well as geometrical and kinematic relationships. Workpiece hardness can be determined using a Vickers hardness test, while the other values can be calculated. The geometrical relationship between grain and workpiece is represented by the maximal uncut chip width ($b_{cu,max}$), the maximal uncut chip thickness ($h_{cu,max}$) and the maximal uncut chip area ($A_{cu,max}$), whereas the kinematic relationship is represented by the grain attack angle α (Fig. 4.13).

As shown in Figs. 4.10 and 4.13, the assumed octahedrally shaped grain results at the end of the chip formation process to a cross-sectional chip surface area of $A_{cu,max}$. The area can be described with:

$$A_{cu,max} = \frac{h_{cu,max} \cdot b_{cu,max}}{2} \tag{4.18}$$

The maximal uncut chip thickness is based on cutting depth (a_e), workpiece speed (v_w), grinding speed (v_c), contact length (l_g), number of grains (N_g) and ratio between uncut chip width and uncut thickness (r). The $h_{cu,max}$ for a triangular shape can be determined with Eq. (4.19) (Rowe 2009, pp. 305):

$$h_{cu,max} = \left(\frac{6 \cdot a_e \cdot v_w}{N_g \cdot r \cdot v_c \cdot l_g}\right)^{\frac{1}{2}} \tag{4.19}$$

In accordance with Rowe, the maximal uncut chip width is defined as shown in Eq. (4.20), where r is typically in the order of 8–15 (Rowe 2009, pp. 305).

$$b_{\mathrm{cu,max}} = h_{\mathrm{cu,max}} \cdot r \tag{4.20}$$

The attack angle (α) can be determined based on grain tip radius (r_g) and maximal uncut chip thickness as follows:

$$\alpha = \arccos\left(\frac{r_g - h_{\mathrm{cu,max}}}{r_g}\right) \tag{4.21}$$

Studies by Challen and Oxley, Xi and Williams as well as Hokkirigawa et al. showed that the level of the attack angle influences the mechanisms of abrasive interaction (rubbing, ploughing or cutting) between grain and workpiece surface (Challen and Oxley 1979, pp. 233; Xie and Williams 1996, pp. 23; Hokkirigawa et al. 1988, pp. 247). Hokkirigawa et al. concluded that each region of the abrasive interaction mechanism can be divided by a theoretical curve as a function of the critical attack angle (Hokkirigawa et al. 1998, pp. 200) (Fig. 4.14).

Figure 4.14 is based on empirical investigations by Hokkirigawa et al. and shows the aforementioned regions of abrasive interaction in dependence with the attack angle and the interfacial coefficient of friction f_{cf}. The interfacial friction coefficient was first introduced by Challen and Oxley and describes the ratio of the strength of a film to the shear flow stress of a soft material (Challen and Oxley 1979, pp. 233). A low value describes low adhesion of the friction partners and a high value a high adhesion. The factor can be used to consider the lubricity of the cutting fluid. Badger and Torrance reported that the interfacial coefficient of friction lies at 0.1 for a neat oil cutting fluid and at 0.4 for a soluble oil cutting fluid (Badger and Torrance 2000, pp. 1111). This results in a slight increase of the necessary critical attack angle to enable cutting instead of ploughing.

Xie and Williams as well as Challen and Oxley also used the interfacial friction coefficient to formulate the friction coefficient μ (Xie and Williams 1996, pp. 23;

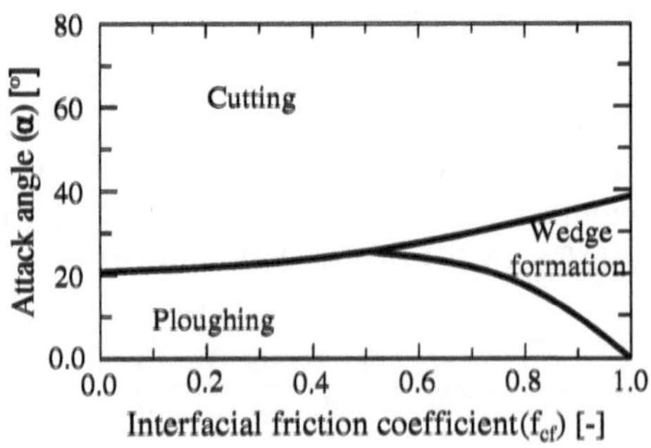

Fig. 4.14 Influence of the interfacial coefficient of friction on the abrasive interaction mechanism (Hokkirigawa et al. 1998, pp. 198)

Challen and Oxley 1979, pp. 237). Investigations by Badger and Torrance showed that the formulation of Xie and Williams for the friction in the micro-cutting mode yielded more accurate results (Badger and Torrance 2000, pp. 1114). Therefore, the formulation by Xie and Williams is also applied in this investigation (see Eq. 4.22). In addition to the mentioned interfacial friction coefficient, the equation also includes the attack angle (α) and the dimensionless distance between adjacent tracks (l), which is given by the authors with 0.2 (Xie and Williams 1996, pp. 23).

$$\mu = \left(\frac{2}{\pi}\right)^{\frac{1}{2}} \cdot \frac{\tan\alpha}{l^{1/4}} \cdot \left(1 - f_{\mathrm{cf}} \cdot \left(1 + \frac{\pi}{4 \cdot \tan^2\alpha}\right)^{\frac{1}{2}}\right) \tag{4.22}$$

Based on Eqs. (4.3) to (4.22), the specific cutting power as well as the specific normal and tangential force can be calculated by looking at the influence of the grinding wheel's characteristics (grain type, size, concentration), workpiece hardness and the cutting fluids lubricity in relation to the grinding and conditioning process parameters.

Abstraction of the real world into an empirical model

Based on existing empirical models of the grinding process presented in Sect. 2.2, a generic nonlinear exponential regression model is chosen (see Eq. 4.23), based on the general regression model presented in Eq. (2.8).

$$Y = K_i + C_i \cdot x_1^{E_1} \cdot x_2^{E_2} \cdots x_n^{E_n} \tag{4.23}$$

The output indicator (Y) is formulated according to the constants (K_i) and (C_i), the input variables (x_i) and the regression coefficients (E_i). The specific normal force and specific tangential force can be modelled using the previously introduced geometrical and kinematical relationships of the grinding and conditioning processes. An example for the specific normal force is presented in Eq. (4.24).

$$F'_{\mathrm{n}} = K_{F'_{\mathrm{n}}} + C_{F'_{\mathrm{n}}} \cdot v_{\mathrm{c}}^{E_{F'_{\mathrm{n},1}}} \cdot v_{\mathrm{w}}^{E_{F'_{\mathrm{n},2}}} \cdot a_{\mathrm{e}}^{E_{F'_{\mathrm{n},3}}} \cdot v_{\mathrm{fad}}^{E_{F'_{\mathrm{n},4}}} \cdot V'^{E_{F'_{\mathrm{n},5}}}_{\mathrm{w}} \cdot v_d^{E_{F'_{\mathrm{n},6}}} \cdot d_{eq}^{E_{F'_{\mathrm{n},7}}} \tag{4.24}$$

Verification of the physical and empirical model

To verify the physical and empirical force models, the grinding process presented in Fig. 4.9 is used to compare the experimental and the calculated results.

For the physical calculation to be completed, it is necessary to specify certain characteristics pertaining to grain, bond, chips, cutting fluid and dresser. The grain tip's radius is defined with 13 μm following Rabiey (2010, pp. 127) and the grain fracture strength equals 2.34 GPa based on investigations by Ichida et al. (2010, pp. 1006). Bond hardness is defined in accordance with Torrance and Badger as $F_0 = 48$ N (Torrance and Badger 2000, pp. 1808). Grain concentration is $K_{\mathrm{g}} = 50\,\%$. The ratio between uncut chip width and uncut thickness is assumed at

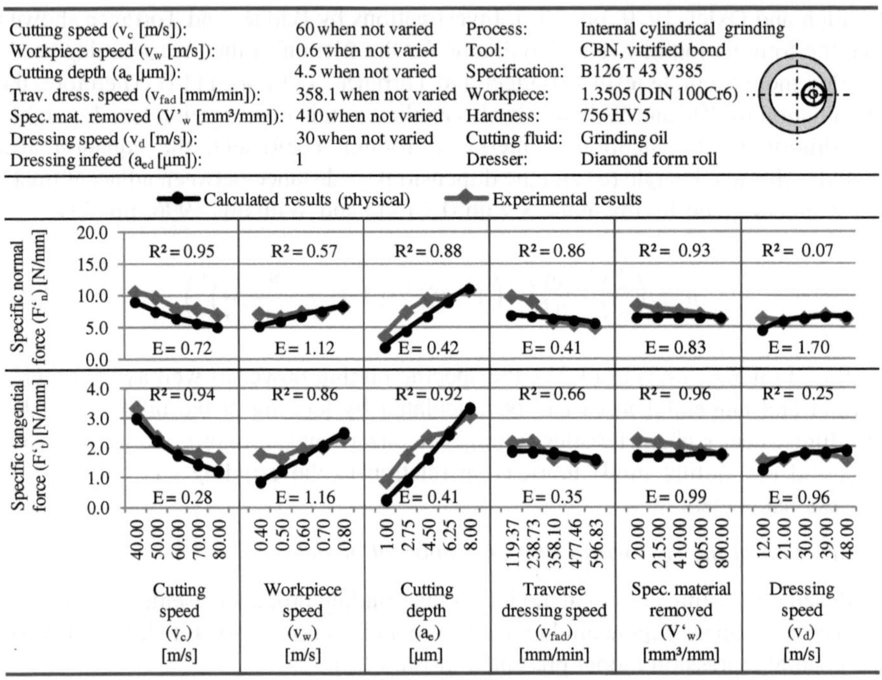

Fig. 4.15 Comparison between experimentally and physically calculated results of the specific forces

12.5. The impact of the applied cutting fluid on the interface friction coefficient is defined with $f = 0.20$. The radius of the diamond form roll is 1 mm.

The results of the physically modelled grinding process are presented in Fig. 4.15. The upper chart shows the comparison for specific normal force and the lower chart for specific tangential force, respectively.

The results are an accurate representation of the influence of the process and its conditioning parameters. With the increase of workpiece speed and cutting depth, the calculated values tend towards a similar result as the experimentally observed values. Similar results apply for cutting speed and traverse dressing speed. The increase of the specific volume of material removed has no influence on the calculated values, as can be seen in the case of the experimental values, because the effect of wear is not considered. Detail analysis, conducted using R^2 and E values, shows an overall good level of accuracy as well. In some cases, however, the R^2 value leads to confusing results. When varying the dressing speed, the R^2 for all values leads to very low or no accuracy, yet the trends seem to fit very well. The deviation can be explained as being the result of the small number of observed values in connection with the stochastic nature of the grinding process. Marginal deviations already have a high impact. In case of dressing speed, for instance, the first and the last observed values show an opposite behaviour to that of the other

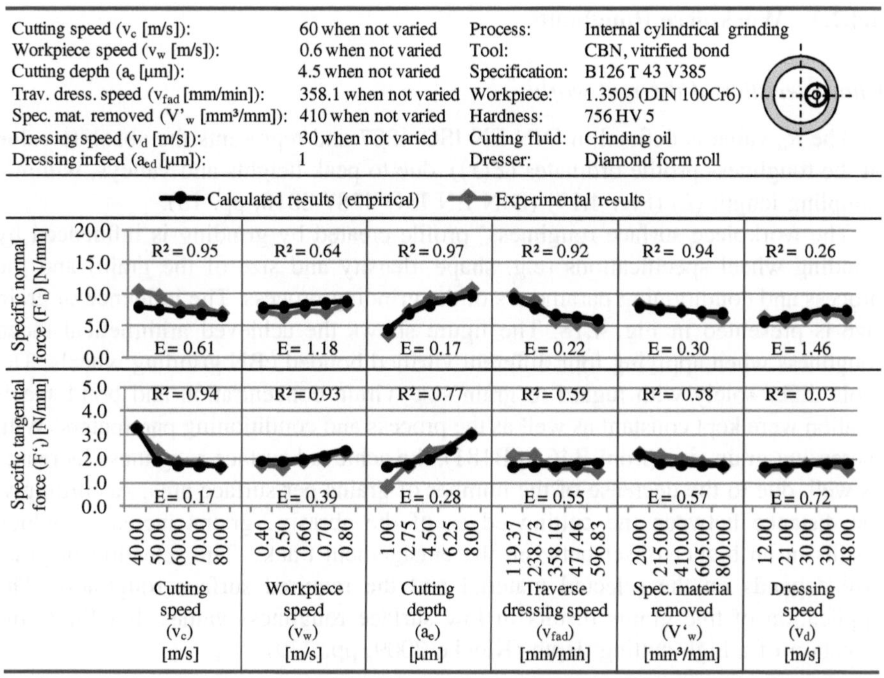

Fig. 4.16 Comparison between experimentally and empirically calculated results for the specific forces

values. If these values were to be changed, according to the assumed trend, then the R^2 value would increase. Therefore, the E value needs to be taken into account to evaluate, if a significant deviation exists. In the discussed cases, the E values are good indicators of accuracy.

Figure 4.16 shows the empirically calculated results in comparison with the previously presented ones. The corresponding regression parameters are shown in Appendix A.1. The first analysis indicates a very good level of accuracy between the empirically calculated values and the observed ones. This is confirmed by looking at the R^2 value and the E value. In case of the dressing speed, however, prediction accuracy is low. This is also due to the small number of observed values in connection with the stochastic process nature. Compared to the R^2 and E values in Fig. 4.15, however, the values in Fig. 4.16 are marginally better.

A general comparison of the physical and empirical models is presented in Sect. 4.4.2.7.

4.4.2.3 Workpiece Roughness

Characteristics of the real world

The R_a value is defined in DIN EN ISO 4287 and represents the absolute value of the roughness profile ordinates ($Z(x)$), due to peak heights and valleys, within a sampling length (L) (Fig. 4.17) (DIN EN ISO 4287 2010, pp. 13).

The workpiece surface roughness' profile created by grinding is influenced by grinding wheel specifications (e.g. shape, density and size of the grain) and the process and conditioning parameters of the grinding process. The influence of grain size is presented in Fig. 4.18. The figure shows the achieved arithmetical mean roughness when applying four different vitrified bonded cBN grinding wheels. The tools differ solely with regard to grain size. Grain concentration and bond specification were kept constant as well as the process and conditioning parameters. With increasing grain size (from B46 to B181), the achieved surface roughness increases as well, due to the decrease of the number of grains per surface area. Accordingly, the distance between the cutting edges of the abrasive grains increases, which results in an higher offset between the engagement tracks. The selection of grain size depends on the selected material and the required surface roughness. The application of fine grains results in low surface roughness values, but limits the selection of a high cutting depth (Klocke 2009, pp. 270).

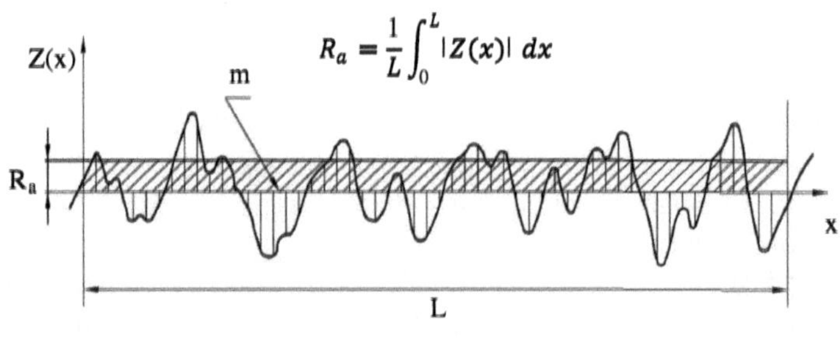

Fig. 4.17 Definition of the arithmetical mean roughness (according to DIN EN ISO 4287 2010, pp. 13)

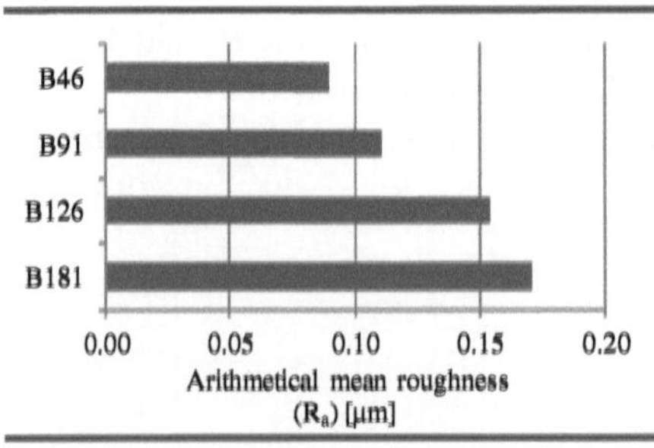

Process:	Internal cylindrical grinding
Tool:	CBN, vitrified bond
Specification:	B46/B91 B126/B181 T 43 V385
Workpiece:	1.3505 (DIN 100Cr6)
Hardness:	756 HV 5
Cutting fluid:	Grinding oil
Dresser:	Diamond formroll
Cutting speed (v_c):	60 m/s
Workpiece speed (v_w):	0.6 m/s
Cutting depth (a_e):	1.7µm
Trav. dress. speed (v_{fad}):	453mm/min
Spec. mat. removed (V'_w):	75 mm³/mm
Dressing speed (v_d):	48m/s
Dressing infeed (a_{ed}):	2µm

Fig. 4.18 Relationship between grain size and arithmetical mean roughness

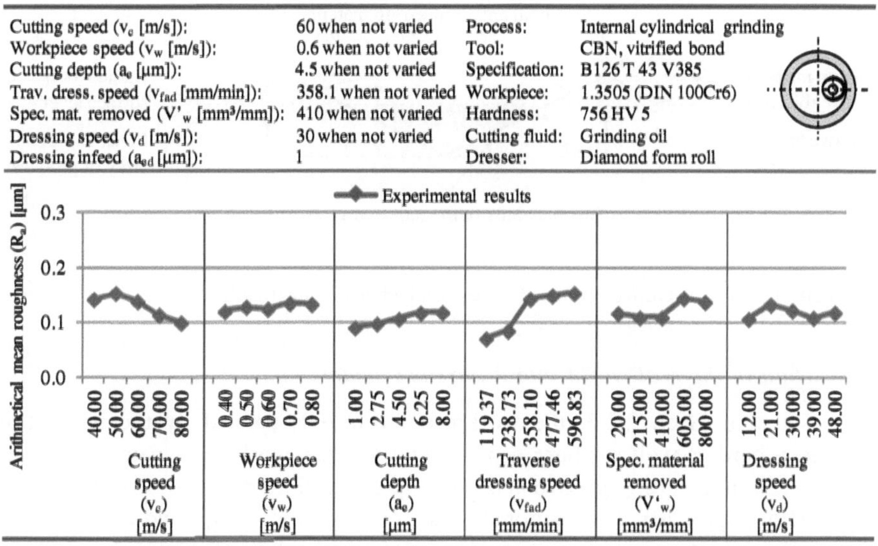

Fig. 4.19 Relationship between process and conditioning parameters and arithmetical mean roughness

The exemplary impact of the process and conditioning parameters on arithmetical mean roughness is presented in Fig. 4.19. The figure presents the measured surface roughness and their dependence on varying settings of v_c, v_w, a_e, v_{fad} and V'_w. The increase of cutting speeds leads to an overall decrease of measured R_a value. This is due to the reason that with increasing cutting speed the cutting depth decreases, if the ratio between workpiece speed and grinding wheel speed is constant. The lower cutting depth leads to a reduction of the forces per grain and overall grinding forces. Lower forces cause lower grinding wheel wear and lower surface roughness (Kassen 1969, pp. 111). An increase of the workpiece speed as well as cutting depth results in an increase of the achieved R_a value, where the influence of the cutting depth is higher than the influence of the workpiece speed. This is due to the previously presented relationships related to cutting forces per grain and the grain wear. Grains splinter faster or break out of the bond, resulting in the engagement of few new and sharp grains which create deeper grooves on the workpiece surface. The rise of the v_{fad} as well as V'_w leads to an increase of the measured R_a value. In case of the v_{fad}, a low value results in a large number of cutting edges produced by conditioning, due to the longer contact time between grain and dresser. A higher number of cutting edges results in a lower surface roughness (Klocke 2009, pp. 155ff.). An increased specific volume of material removed results in an increase of contact time, the grain wear and subsequently the surface roughness value.

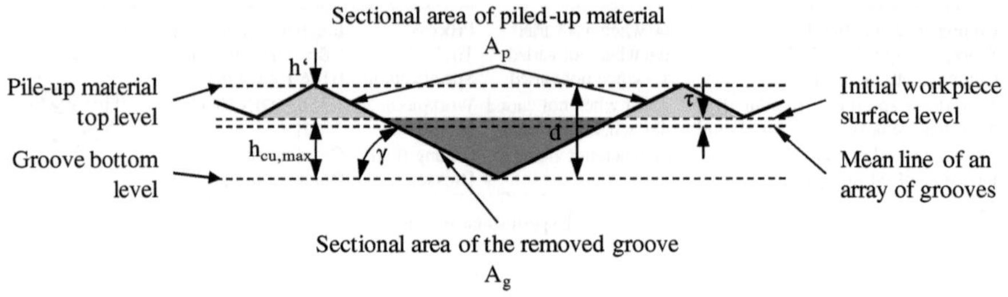

Fig. 4.20 Theoretical groove produced by an octahedrally shaped grain

Abstraction of the real world into a physical model

As stated in Sect. 2.2.3, the workpiece surface is created by the interaction between a hard as well as sharp asperity and a softer workpiece material. During this interaction, a cutting groove and piled up material on both sides of the groove are created (Fig. 4.20). The sectional area's cutting groove (A_g) and piled-up (A_p) material can be described based on the geometrical relationships of the angle of piled-up material (γ), the maximal undeformed chip thickness ($h_{cu,max}$), the height of the piled-up material (h'), the distance between the mean line of an array of grooves and the initial workpiece surface level (τ) and groove depth (d_{groove}).

In addition to the assumptions presented for the modelling of the specific cutting forces, the following assumptions apply as well:

- grains have an octahedronal shape, which will create triangular grooves on the workpiece surface (Torrance and Badger 2000, pp. 1795),
- piled-up material is created on both sides of the groove and has a symmetrical form (Chen and Rowe 1996, pp. 893) and
- the area of the material removed is proportional to the maximal uncut chip area of the grain (Chen and Rowe 1996, pp. 888).

Considering the geometrical relationships presented in Fig. 4.20 **as well as the cutting efficiency** (e), the arithmetical mean roughness can be formulated as follows (Torrance and Badger 2000, pp. 1796):

$$R_a = \left(\frac{h' - \tau}{2}\right) \cdot \sqrt{\frac{e \cdot d_{groove}}{2 \cdot (h' - \tau)}} \tag{4.25}$$

The first fraction in Eq. (4.25) describes the surface roughness created by an array of grooves. The random superposition of the groove is represented via the multiplication with the later fraction. The knowledge of cutting efficiency, however, is necessary for the calculation. Cutting efficiency describes the ratio between the volume of the material removed and the volume piled-up on both sides of the groove. Torrance and Badger define the cutting efficiency Eq. (4.26), based on abrasive wear coefficients proposed by Xie and Williams.

$$e = 1 - \left(\frac{K_c}{K_{c,max}} \right) \tag{4.26}$$

Xie and Williams formulated the abrasive wear coefficients (K_c) based on experimental measurements (Xie and Williams 1996, pp. 24). On this basis, Torrance and Badger modified the original equation (Torrance and Badger 2000, pp. 1792):

$$K_c = 0.003 \cdot \frac{27 \cdot (I \cdot \tan \alpha)^3}{f_{cf} \cdot H_{s,w} \cdot \sqrt{l}} \cdot \sqrt{\frac{H_{b,w}}{H_{s,w}}} \tag{4.27}$$

The equation (Eq. 4.27) is based on the attack angle (α), the interfacial friction coefficient (f_{cf}), the dimensionless distance between adjacent tracks (l), a statistical factor (I), the hardness of the bulk material ($H_{b,w}$) and of the workpiece surface ($H_{s,w}$). For the statistical factor, Torrance and Badger report a value of $I = 1.185$.

The maximum abrasive wear coefficient ($K_{c,max}$) is described by Torrance and Badger as the ratio of attack angle (α) and the workpiece surface hardness ($H_{s,w}$) (Torrance and Badger 2000, pp. 1792):

$$K_{c,max} = \frac{\tan \alpha}{H_{s,w}} \tag{4.28}$$

After having obtained cutting efficiency, the sectional area of piled-up material (A_p) can be determined. The basis is the sectional area of the cutting groove (A_g), which equals the maximal uncut chip area ($A_g = A_{cu,max}$) (Chen and Rowe 1996, pp. 888):

$$A_p = \frac{((1 - e) \cdot A_g)}{2} \tag{4.29}$$

Owing to the geometrical relationships, the height of the piled-up material (h') can be formulated according to A_p and the angle of piled-up material (γ):

$$h' = \sqrt{A_p \cdot \tan \gamma} \tag{4.30}$$

with

$$\tan \gamma = \frac{b_{cu,max}}{2 \cdot h_{cu,max}} \tag{4.31}$$

The groove depth (d_{groove}) can then be derived by the sum of the h' and $h_{cu,max}$ (Eq. 4.32).

$$d_{\mathrm{groove}} = h' + h_{\mathrm{cu,max}} \tag{4.32}$$

To calculate the distance between the mean line of an array of grooves and the initial workpiece surface level (τ), Torrance and Badger proposed the following equation based on d_{groove} and h' (Torrance and Badger 2000, pp. 1809):

$$\tau = \frac{h'^2 - d_{\mathrm{groove}}^2 - 2 \cdot d_{\mathrm{groove}} \cdot h'}{2 \cdot d_{\mathrm{groove}} \cdot h'} \tag{4.33}$$

However, the actual formulation of the model used to determine the R_a value lacks focus on conditioning parameters. Accordingly, investigations were conducted, taking into consideration the weighting factor (f_{U_d}) based on the dimensionless dressing overlap ratio (U_d) (see Eq. 2.5). The adjusted arithmetical mean roughness (R_a^*) can be described:

$$R_\mathrm{a}^* = R_\mathrm{a} \cdot f_{U_\mathrm{d}} \tag{4.34}$$

With the weighting factor formulated as:

$$f_{U_\mathrm{d}} = \left(1 + \frac{1}{U_\mathrm{d}}\right)^2 \tag{4.35}$$

The arithmetical mean roughness can be described in connection with the process and conditioning parameters. Furthermore, the equations consider the lubrication capacity of the cutting fluid and the influence of the grinding wheel's specifications.

Abstraction of the real world into an empirical model

The empirical model used to determine the arithmetical mean roughness is based on the general regression model presented in Eq. (2.8). The model is formulated on the geometrical and kinematical relationships of the grinding and the conditioning processes (see Eq. 4.36).

$$R_\mathrm{a} = K_{R_\mathrm{a}} + C_{R_\mathrm{a}} \cdot v_\mathrm{c}^{E_{R_\mathrm{a},1}} \cdot v_\mathrm{w}^{E_{R_\mathrm{a},2}} \cdot a_\mathrm{e}^{E_{R_\mathrm{a},3}} \cdot v_\mathrm{fad}^{E_{R_\mathrm{a},4}} \cdot V{\prime}_\mathrm{w}^{E_{R_\mathrm{a},5}} \cdot v_\mathrm{d}^{E_{R_\mathrm{a},6}} \cdot d_\mathrm{eq}^{E_{R_\mathrm{a},7}} \tag{4.36}$$

Verification of the physical and empirical model

The arithmetical mean roughness is calculated for the grinding case presented in Fig. 4.19 by applying the equations to the physical and empirical modelling. Figure 4.21 shows the result of the comparison between observed values and calculated values. Charts A and B show the extent to which the physical models fit with the observed values and chart C fits with empirically determined values. The chart A presents the result for the R_a value according to Eq. (4.25) and the chart B shows the result for the adjusted R_a^* value based on Eq. (4.34).

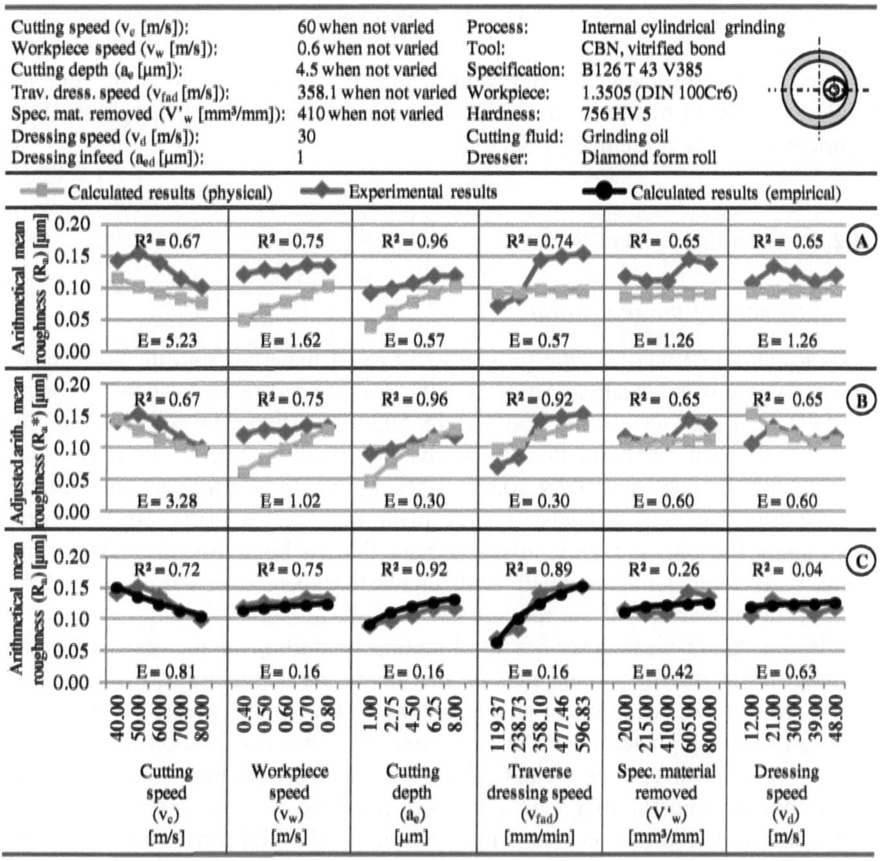

Fig. 4.21 Comparison of experimental and calculated (physical and empirical) results of the arithmetical mean roughness

The physical calculations are based on the aforementioned characteristics for grain, bond, chips, cutting fluid and dresser (see verification of specific cutting forces). Workpiece bulk hardness was measured with 756 MPa. The regression parameters of the empirical model are presented in Appendix A.1.

At a first glance, the results of A and B indicate an overall good level of accuracy and chart C a very good level of accuracy in the matching of the calculated and measured results. The calculated and measured R_a values presented in charts A and B show a similar behaviour when increasing or decreasing process and conditioning parameters. The gradient of measured and calculated curves, however, differs in the case of workpiece speed and cutting depth, while cutting speed and traverse dressing speed show a comparable gradient. By including the weighting factor to account for the influence of the conditioning parameter, the matching between calculated and measured values of chart B can also be increased in comparison with chart A. The consideration of the coefficient of determination and the relative absolute error between measured and calculated values shows a moderate accuracy. When comparing the three charts, chart C is the most accurate in its matching of observed and empirically derived values, due to comparably high R^2 and low E values, followed by charts B and A. A general comparison of the physical and empirical model is presented in Sect. 4.4.2.7.

4.4.2.4 Grinding Wheel Wear

Characteristics of the real world

Wear can be defined as "the progressive loss of material from the surface of a solid body due to mechanical action, i.e. the contact and relative motion against a solid, liquid or gaseous counter body" (DIN 50320, 1979, pp. 1). According to DIN 50320, four wear phenomena can be distinguished: adhesive, abrasive, tribo-chemical and fatigue wear (DIN 50320, 1979, pp. 6). In grinding, a super-position of these four phenomena leads to three main wear mechanisms: attritious grain wear, grain fracture and grain dislodgement due to bond fracture. These three mechanisms occur simultaneously with increasing specific material removal rate resulting in an increase of the radial wear (Fig. 4.22).

Attritious grain wear describes the blunting or dulling of the abrasive grains, resulting in the growth of wear flats on the tips of the active grains (Malkin 1968, pp. 36). Micro-wear occurs due to the rubbing of the grains against the workpiece surface, causing high temperatures and pressure-induced oxidation and diffusion processes on the surfaces in contact. With increasing specific material removal rate, the mechanical and thermal load increases, resulting in slow grain fatigue. The fatigue leads, in connection with imperfections of the grain material, to a partial fracturing of grain into abrasive particles. Grain dislodgement results from the fracturing and erosion of the bond post, due to a mechanical overload at high

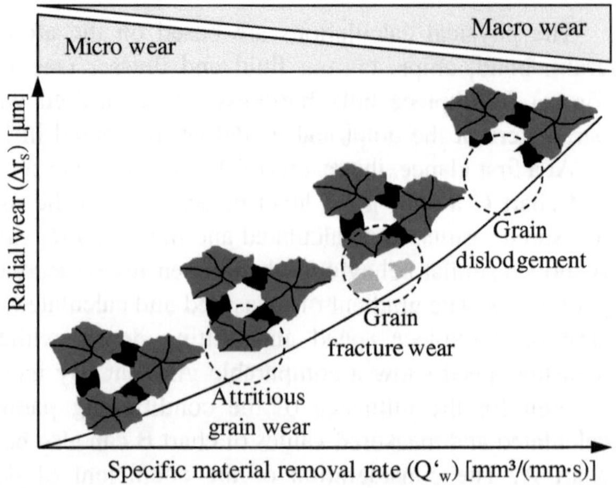

Fig. 4.22 Grain and bond wear mechanism in grinding (Bierlich 1976, pp. 76)

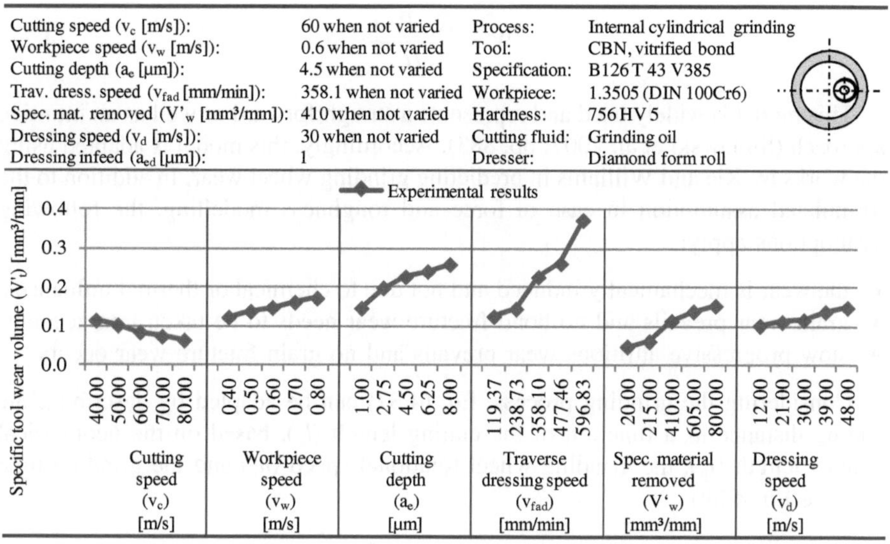

Fig. 4.23 Relationship between process and conditioning parameters and the specific grain wear volume

specific material removal rates (Grüntzig 2011, pp. 15). A typical wear pattern is the attritious wear-induced grain blunting, followed by the creation of sharper grains due to grain fracturing until the bond posts fracture and the grains dislodge. During this process, attritious grain wear has the smallest contribution to the overall radial wear. Wheel life, however, is determined by the attritious wear, because wear reduced grain sharpness and protrusion height leads to wheel glazing (Fathima et al. 2003, pp. 1247).

The influence of process and conditioning parameters on specific tool wear volume is presented in Fig. 4.23. With increasing cutting speed, the wear volume declines, due to the aforementioned reduction of the cutting depth (if the ratio between workpiece speed and grinding wheel speed is constant). Accordingly, the load on the grain and the wear declines.

The opposite wear behaviour occurs when workpiece speed and cutting depth are increased, due to the increased grain load. As mentioned before, the increase of the traverse dressing speed leads to a reduction in the number of cutting edges. The load on each grain increases and the wear increases as well. With increasing V'_w, the contact length or contact time between grain and workpiece increases as well and grain wear increases.

Abstraction of the real world into a physical model

To provide a general and simple model, Archard proposed a prediction model of the wear volume (W) as a function of sliding distance (s), wear coefficient (K), applied normal load (P) and material hardness (H) (see Eq. 4.37) (Archard 1961, pp. 1420ff.).

$$W = \frac{K \cdot P \cdot s}{H} \tag{4.37}$$

This model is widely used and represents a straightforward as well as an intuitive approach (Siniawski et al. 2007, pp. 883). Accordingly, this model is adapted using the works by Xie and Williams in predicting grinding wheel wear. In addition to the formulated assumption in case of force and roughness modelling, the following assumptions apply:

- the wear is mechanically induced and not due to chemical or thermal influences;
- grain wear prevails and no bond fracture wear needs to be taken into account;
- slow progressive attritious wear prevails and no grain fracture wear occurs.

Considering the grinding process, Eq. (4.37) can be adapted by describing the sliding distance as a function of the cutting length (l_c), based on the geometrical contact length (l_g), the grinding wheel rotational speed (n_s) and the grinding time (t_c) (see Eq. 4.38).

$$s = l_c = l_g \cdot n_s \cdot t_c \tag{4.38}$$

The dimensionless wear coefficient (K) can be formulated as the product of the dimensional abrasive wear coefficient of the grain ($K_{c,g}$) and grain surface hardness ($H_{s,g}$) (see Eq. 4.39). $K_{c,g}$ can be determined by including grain surface ($H_{s,g}$) and bulk material ($H_{b,g}$) hardness into Eq. (4.27) instead of the workpiece surface and bulk material hardness.

$$K = K_{c,g} \cdot H_{s,g} \tag{4.39}$$

By including the normal force (F_n), based on Eq. (4.11), instead of the normal load (P), the tool wear volume (V_t) can be formulated based on Eqs. (4.37), (4.38) and (4.39) to:

$$V_t = K_{c,g} \cdot F_n \cdot l_c \tag{4.40}$$

To determine the specific tool wear volume (V'_t), Eq. (4.40) can be related to 1 mm grinding width (a_p):

$$V'_t = \frac{K_{c,g} \cdot F_n \cdot l_c}{a_p} \tag{4.41}$$

In a similar fashion to Eq. (4.25), Eq. (4.41) lacks the consideration of conditioning parameters. Therefore, an adjusted specific tool wear volume can be formulated (V'^{*}_t) (see Eq. 4.42), taking the conditioning process weighting factor (f_{Ud}) into account.

$$V_{\mathrm{t}}^{\prime *} = \frac{K_{\mathrm{c,g}} \cdot F_{\mathrm{n}} \cdot l_{\mathrm{c}}}{a_{\mathrm{p}}} \cdot f_{U_{\mathrm{d}}} \tag{4.42}$$

Abstraction of the real world into an empirical model

In order to model the specific tool wear volume empirically, the formulated equation (Eq. 4.43) is based on the general regression model presented in Eq. (2.8). The equation includes geometrical and kinematical relationships of the grinding and the conditioning processes.

$$V'_{\mathrm{t}} = K_{V'_{\mathrm{t}}} + C_{V'_{\mathrm{t}}} \cdot v_{\mathrm{c}}^{E_{V'_{\mathrm{t},1}}} \cdot v_{\mathrm{w}}^{E_{V'_{\mathrm{t},2}}} \cdot a_{\mathrm{e}}^{E_{V'_{\mathrm{t},3}}} \cdot v_{\mathrm{fad}}^{E_{V'_{\mathrm{t},4}}} \cdot v_{d}^{E_{V'_{\mathrm{t},6}}} \cdot V_{\mathrm{w}}^{\prime E_{V'_{\mathrm{t},5}}} \cdot d_{\mathrm{eq}}^{E_{V'_{\mathrm{t},7}}} \tag{4.43}$$

Verification of the physical and empirical model

Figure 4.24 shows a comparison of the observed and predicted values in the equation of physical and empirical models. Chart A presents specific tool wear volume, chart B adjusted specific tool wear volume and chart C the empirically derived roughness according to process and conditioning parameters. The cBN's grain bulk hardness was assumed to be 45 GPa (Davis 1995, pp. 109). The normal force was calculated based on Eq. (4.11). The parameters for regression are presented in the Appendix A.1.

In a first analysis, charts A and B in particular show a below average level of accuracy, while chart C shows a good match between observed and calculated values. The deviation between observed and calculated values actually increases when taking into account the conditioning process weighting factor. The only advantage when considering the weighting factor is the corresponding gradient in case of traverse dressing speed; otherwise, its use leads to a decrease in accuracy. This analysis is verified by focussing on the E value for charts A and B. The values of charts A and B are significantly higher than the E values for chart C, which are comparatively lower. The R^2 values are also considerably higher for chart C, indicating an overall good level of accuracy. In case of charts A and B, the R^2 value confirms that there is a good match between measured and predicted values.

A general comparison of the physical and empirical models is presented in Sect. 4.4.2.7.

4.4.2.5 Power Demand

Characteristics of the real world

As presented in Fig. 2.10, the realisation of chip formation, the operation of components, the application of compressed air or the cooling of cutting fluid

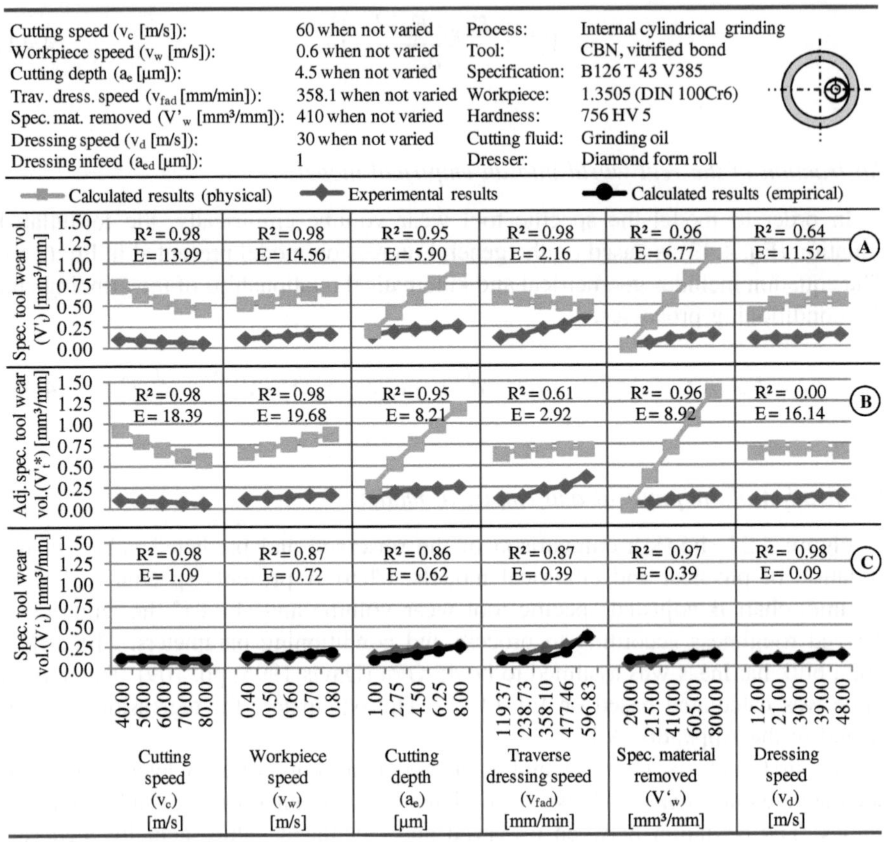

Fig. 4.24 Comparison between experimental and calculated (physical and empirical) results of specific tool wear volume

generates a demand for electrical, thermal and mechanical power. While power demand throughout the grinding process depends on the selected conditioning and process parameters, the power demand of the grinding system depends on the temperature as well as the operational state of the components and compressed air valves. Corresponding components are the drives, pumps, cooling units, control units, actuators, signal elements, wiring and measuring systems (Zein 2012, pp. 11). The compressed air valves control the application of compressed air as spindle sealing and purging air, control air (e.g. valves of doors, tool clamping, workpiece position) as well as cleaning air of tools and workpieces (Kuhrke 2011, pp. 39). The demand of power depends on the size and number of the components and valves as well as their operational condition, leading to huge variations between minimum and maximum values for power demand. According to the Manufacturing Technologies Association (United Kingdom), the power rating of grinding

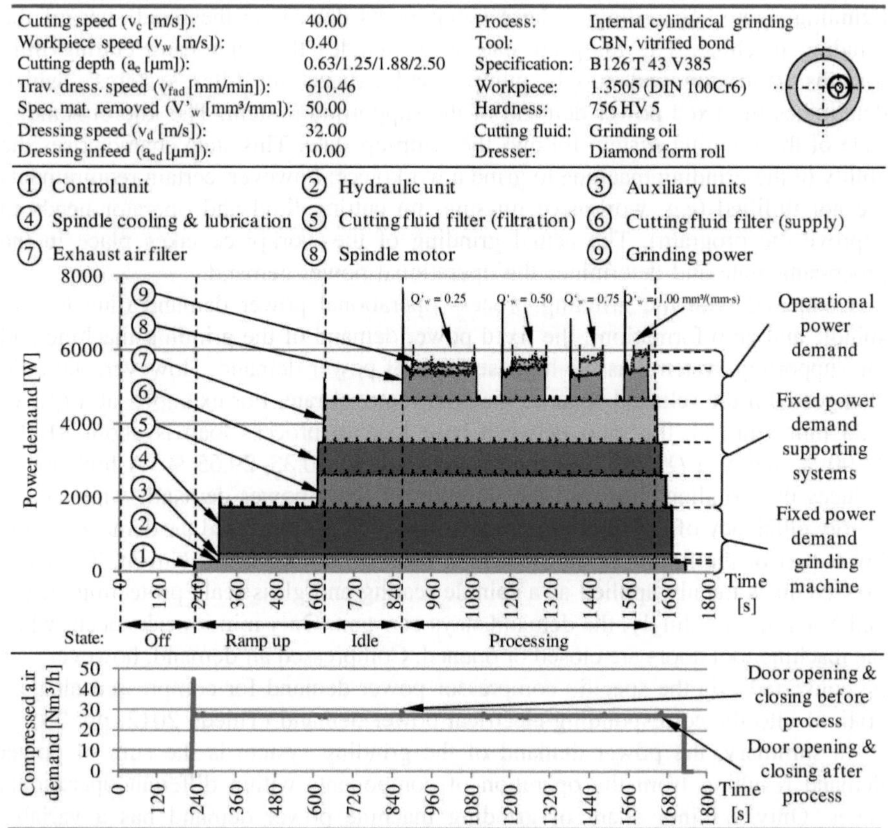

Fig. 4.25 Power demand area chart of a grinding system when performing an internal cylindrical grinding process

machines can range between a minimum value of about 0.25 kW and a maximum value of 50 kW, with a mean power rating of 15 kW (SKM Enviros 2011, pp. 20).

As an example, Fig. 4.25 shows the power demand area chart of an internal cylindrical grinding process. The area chart includes, in the upper part, the power demand of the grinding machine (control unit, spindle cooling and lubrication, hydraulic unit and spindle motor), supporting systems (due to cutting fluid and exhaust air filter) and the grinding process (considering four different specific material removal rates) as well in the lower part, compressed air demand.

The figure also shows the distribution of fixed and operational power demand (Li et al. 2011, pp. 269). The classification of the distribution is the result of different operational states (off, ramp up, idle, processing) (Thiede 2012, pp. 100). In the ramp-up state, the components are progressively turned on. In case of the grinding machine, ramp up starts with the control unit followed by the hydraulic unit, spindle cooling and lubrication unit as well as further auxiliary units (e.g.

lightning, fans). Accordingly, these components determine the fixed state of the grinding machine. The ramp-up state also includes the supporting cutting fluid systems' (filtration and supply pumps) and exhaust air filter systems', which determines the fixed power demand of the supporting system. The idle or standby state of the grinding system follows the ramp-up state. This state characterises the ability of the grinding machine to grind a workpiece; however, certain requirements are not fulfilled (e.g. workpiece missing, no cutting fluid and operator needs to approve the program). The actual grinding of the workpiece takes place in the processing state and determines the operational power demand.

Compared with the grinding process operational power demand (due to tool spindle and chip formation), the fixed power demand of the grinding machine and the supporting system has the highest share of power demand. However, the ratio changes with the selected specific material removal rate. For example, at a Q'_w of 0.25 $mm^3/(mm\ s)$, the ratio between base load to process load is about 81.10–18.90 % and at a Q'_w of 1.00 $mm^3/(mm\ s)$ about 70.35–29.65 %. A higher Q'_w reduces the grinding time and the duration of fixed power demand, whereas the energy efficiency of the grinding process increases. Compressed air demand, in the lower part of the figure, is on a constant level of around 27.20 Nm^3/h. The compressed air is mainly applied as a spindle sealing and glass scale protecting air (*X*- and *Y*-axis); accordingly, the demand stays constant. Only minor peaks occur when the machine tool doors are closed or opened. Compressed air demand, however, can be converted (via the specific compressor power demand for compressed air generation) into the corresponding electrical power demand (Thiede 2012, pp. 28).

In summary, the power demand of the grinding system is the sum of power demand resulting from the operation of components within different operational states. Only a minor share of grinding machine power demand has a variable behaviour, and this share is used for value creation. A major share of power demand is not variable, but is necessary to guarantee the operating ability. This conclusion is also confirmed by investigations by Li et al., Zein and Kuhrke (Li et al. 2011, pp. 270; Zein 2012, pp. 87ff.; Kuhrke 2011, pp. 64ff.). The modelling of power demand needs to consider the variable and fix behaviour of single components as well as influencing parameters.

Abstraction of the real world into physical and empirical models

Based on Figs. 2.8 and 4.25 and on studies by Li et al., Zein and Kuhrke (Li et al. 2011, pp. 270; Zein 2012, pp. 87ff.; Kuhrke 2011, pp. 64ff.), Table 4.1 shows the systems, sub-systems and components considered for modelling. The grinding system includes the sub-systems' grinding machine, cutting fluid filter and exhaust air filter. These three sub-systems can be further structured into components. Additional sub-systems could be the workpiece changer or metring equipment; however, these sub-systems are not considered.

In the table, two operational states are distinguished with the fixed state (F), (characterised by a constant power demand) and the variable state (V) (characterised by a variable power demand). The classification into F or V states is based on investigations by Li et al. and Kuhrke (Li et al. 2011, pp. 270;

Table 4.1 Categorisation of components in sub-systems and systems

System	Subsystem	Components	State
Grinding process	Chip formation	• Tool • Workpiece • Cutting fluid	V V V
Grinding system	Grinding machine	• Spindle drives (tool, workpiece, dresser) • Servo drives • Hydraulic unit • Control unit • Auxiliary units (e.g. lightning, fans) and spindle cooling and lubrication unit	V V F F F F
	Exhaust air filter	• Ventilator drives and • Filtration unit	F F
	Cutting fluid filter	• Filtration units,	F/V[a]
		• Supply units and	F/V[a]
		• Cooling units (optional)	F

V variable operational state
F fixed operational state
[a] Dependent on used drives and pumps as well as set-up

Kuhrke 2011, pp. 64ff.). As far as cutting fluid filter is concerned, two states are possible for filtration and supply units due to the application of controlled or uncontrolled pumps (Kuhrke 2011, pp. 88). Uncontrolled pumps deliver a constant cutting fluid volume flow at a constant fluid pressure, resulting in a fixed power demand. Controlled pumps (applying variable-frequency drives) can convey the needed volume flow according to the needed fluid pressure with varying drive speeds, resulting in a situation-dependent and usually lower power demand.

To model the power demand of a grinding system, it is important to consider the component characteristics in accordance with the operational state and selected conditioning and process parameters. The following assumptions apply in order to reduce effort and complexity:

- only the power demand of idle and processing state is considered,
- the power demand of components with a fixed operational state is constant and corresponds to the observed values at this state (Li et al. 2011, pp. 270) and
- observed values at a fixed operational state are determined using a standardised measuring procedure (Zein 2012, pp. 81ff.).

The physical and empirical modelling is partially based on studies by Kuhrke (2011, pp. 1ff.) as well as own investigations. The modelling is separated into fixed and the operational power demand. The focus of the next section is on the description of fixed power demand, followed by variable power demand and grinding system power demand.

Modelling of the fixed power demand

When considering the grinding machine, usually four operational states can be distinguished. However, the off as well as ramp-up states can be neglected. In the off state, usually no power demand occurs and the ramp-up phase only occurs when the grinding machine is turned on. Therefore, the modelling of the grinding machine needs to take into account the power demand of the idle and processing phase. As already mentioned, the idle state is characterised by a phase where the machine tool is ready to produce. Power demand ($P_{\mathrm{GM,idle}}$) in this state can be described as the sum of the hydraulic unit (P_{hu}), control unit (P_{cu}), spindle cooling and lubrication unit (P_{scl}) and auxiliary units (P_{au}) (see Eq. 4.44).

$$P_{\mathrm{GM,idle}} = P_{\mathrm{hu}} + P_{\mathrm{cu}} + P_{\mathrm{scl}} + P_{\mathrm{au}} \tag{4.44}$$

The grinding process starts during the processing state, which, in addition to $P_{\mathrm{GM,idle}}$, requires more power for chip formation/cutting (P_{c}), grinding wheel ($P_{\mathrm{s,gw}}$) and workpiece ($P_{\mathrm{s,w}}$) spindle resulting in $P_{\mathrm{GM,proc,g}}$.

$$P_{\mathrm{GM,proc,g}} = P_{\mathrm{GM,idle}} + P_{\mathrm{c}} + P_{\mathrm{s,gw}} + P_{\mathrm{s,w}} \tag{4.45}$$

The conditioning process can also be considered as a processing state. The cutting power, however, is negligible and the necessary power demand during dressing ($P_{\mathrm{GM,proc,d}}$) depends on the idle power ($P_{\mathrm{GM,idle}}$) as well as on the spindle power of the grinding wheel ($P_{\mathrm{s,gw}}$) and dresser ($P_{\mathrm{s,d}}$).

$$P_{\mathrm{GM,proc,d}} = P_{\mathrm{GM,idle}} + P_{\mathrm{s,gw}} + P_{\mathrm{s,d}} \tag{4.46}$$

As previously mentioned in Sect. 2.1.3, either a centralised or a decentralised exhaust air and cutting fluid filtration system can be connected to the grinding machine. Therefore, the power demand due to either one of these sub-systems needs to be accounted for. In case of a decentralised exhaust air filtration system, generally two operational states (off and processing/filtering) can be distinguished, due to a short ramp-up time and usually no idle time. In the processing state, cutting fluid aerosols are evacuated from the encapsulated grinding machine interior room. The exhaust air volume flow (Q_{ea}) remains constant (e.g. 500, 1000, 2500 $\mathrm{m^3/h}$). The selection of the needed exhaust air volume flow depends on the size of the interior room, the expected amount of cutting fluid aerosol and the applied cutting fluid. If a water miscible cutting fluid is applied, then the needed exhaust air volume flow is lower compared with the application of non-water miscible cutting fluids (LTA Lufttechnik 2014, pp. 2ff.). The exhaust air filter sub-system can be described based on the power demand of the ventilator drive ($P_{\mathrm{EAF,v}}$) and filtration units ($P_{\mathrm{EAF,f}}$). Due to the extraction of a constant exhaust air volume flow, there is a constant power demand. Both demands can be added up to the fixed power demand in the processing state of the decentralised filter ($P_{\mathrm{EAF,proc}}$).

$$P_{\mathrm{EAF,proc}} = P_{\mathrm{EAF,f}} + P_{\mathrm{EAF,v}} \tag{4.47}$$

If a centralised exhaust air filtration system is used, then the overall power demand ($P_{\mathrm{EAF,proc,total}}$) needs to be related to the ratio of needed (Q_{ea}) and total exhaust air volume flow ($Q_{\mathrm{ea,total}}$), to obtain the grinding system's exhaust air filtration power demand ($P_{\mathrm{EAF,proc}}$).

$$P_{\mathrm{EAF,proc}} = \frac{P_{\mathrm{EAF,proc,total}}}{Q_{\mathrm{ea,total}}} \cdot Q_{\mathrm{ea}} \tag{4.48}$$

In case of the cutting fluid filter, the fixed power demand of the filtration, supply and cooling units is connected with the design of the used pumps or drives. In case of uncontrolled drives, a fixed rotational speed and a fixed power demand arises. Under the assumption that the operation of the pumps results in a constant high heat input into the fluid, then the cutting fluid cooling unit has also a fixed power demand. The operation of the cutting fluid filter also shows the four operational states described in Fig. 4.25. After a short ramp-up state (Fig. 4.25), the idle state can be characterised with the operation of the filtration unit ($P_{\mathrm{CFF,f}}$) and cooling unit ($P_{\mathrm{CFF,cl}}$), resulting in idle power demand ($P_{\mathrm{CFF,idle}}$) (see Eq. 4.49). During the processing state, the cutting fluid supply ($P_{\mathrm{CFF,s}}$) unit and lifting pumps sometimes used ($P_{\mathrm{CFF,l}}$) are activated and determine power demand ($P_{\mathrm{CFF,proc}}$) (see Eq. 4.50).

$$P_{\mathrm{CFF,\,idle}} = P_{\mathrm{CFF,\,f}} + P_{\mathrm{CFF,\,cl}} \tag{4.49}$$

$$P_{\mathrm{CFF,proc}} = P_{\mathrm{CFF,f}} + P_{\mathrm{CFF,s}} + P_{\mathrm{CFF,cl}} + P_{\mathrm{CFF,l}} \tag{4.50}$$

If a centralised cutting fluid filter is used, power demand can be determined in correspondence with Eq. (4.48). The total centralised cutting fluid filter power demand ($P_{\mathrm{CFF,proc,total}}$) is related to the cutting fluid flow (Q_{cf}) needed in the grinding process and total fluid flow of the centralised cutting fluid filter ($Q_{\mathrm{cf,total}}$).

$$P_{\mathrm{CFF,proc}} = \frac{P_{\mathrm{CFF,proc,total}}}{Q_{\mathrm{cf,total}}} \cdot Q_{\mathrm{cf}} \tag{4.51}$$

Power demand related to the compressed air (P_{ca}) can be determined by multiplying the compressed air demand (Q_{ca}) by the specific compressor power demand for compressed air generation (P'_{ca}) (see Eq. 4.52).

$$P_{\mathrm{ca}} = P'_{\mathrm{ca}} \cdot Q_{\mathrm{ca}} \tag{4.52}$$

Modelling of the variable power demand

The following modelling part considers components with a variable operational state. These components are modelled in relation to their influencing variables.

Corresponding components and systems are the grinding process, the spindle drives of the machine tool as well as filtration and supply units of the cutting fluid filter. An exception is made with regard to the modelling of the servo drives. These components have a variable power demand depending on the selected feed. However, their power demand is comparatively lower, due to small feeds during the grinding process, and is not considered. Further information on the empirical modelling of the servo drive power demand is presented by Kuhrke (2011, pp. 79ff.).

The variable power demand of the grinding process (P_c) can be modelled based on the specific tangential force (F'_t) from Eq. (4.12) in dependence with the grinding width (a_p) and the cutting speed (v_c):

$$P_c = F'_t \cdot a_p \cdot v_c \tag{4.53}$$

As mentioned, a variable power demand occurs within the grinding system due to the spindle drives operated to rotate the grinding wheel, the workpiece and the dresser. In order to predict the spindle power demand ($P_{s,i}$), empirical regression models (regression coefficients: $a_{1,i}$, $a_{2,i}$ and b_i) can be applied based on the power change in dependence with rotational speed (n_i) (see Eq. 4.54).

$$P_{s,i} = a_{1,i} \cdot n_i^2 + a_{2,i} \cdot n_i + b_i \tag{4.54}$$

i stands for grinding wheel (gw), workpiece (w) and dresser (d) spindle.

In case of the sub-system cutting fluid filter, the application of controlled drives results in a variable power demand. The power demand is a function of the cutting fluid's volume flow (Q_{cf}), the pressure (p_{cf}) and kinematic viscosity (ν_{cf}) as well as the pump design. Kuhrke presented an empirical approach to determine the power demand of controlled pumps (Kuhrke 2011, S91ff.). The approach determines the power demand according to cutting fluid volume flow and pressure. The approach leaves out cutting fluid viscosity. Therefore, a different approach, using a nonlinear regression to determine the influence of Q_{cf}, p_{cf} and ν_{cf}, is presented in the following paragraphs. The necessary data can be derived from pump performance curves. Within these performance curves, the mechanical power demand (P_{mech}) is provided by the pump manufacturer. Using the general regression model presented in Eq. (2.8), the P_{mech} can be modelled including Q_{cf}, p_{cf} and ν_{cf} (see Eq. 4.55).

$$P_{mech} = K_{CFF} + C_{CFF} \cdot \nu_{cf}^{E_{CFF,1}} \cdot p_{cf}^{E_{CFF,2}} \cdot Q_{cf}^{E_{CFF,3}} \tag{4.55}$$

P_{mech} is connected via the drive efficiency (η_{el}) with the electrical power demand (P_{el}) of the pump ($P_{CFF,f}$, $P_{CFF,s}$ or $P_{CFF,l}$ respectively) (see Eq. 4.56).

$$P_{el} = \frac{P_{mech}}{\eta_{el}} = P_{CFF,i} \tag{4.56}$$

i stands for filtration, supply or lifting (f, s or l).

While the v_{cf} is defined by the applied cutting fluid, p_{CFF} and Q_{cf} are defined by the grinding process. The selection of fluid pressure depends on the needed cutting fluid jet velocity (v_{cf}) when applied into the contact area via the cutting fluid orifice, as well as fluid density (ρ_{cf}) (see Eq. 4.57). A cutting fluid jet velocity which equals the grinding speed is usually opted for.

$$p_{cf} = \frac{\rho_{cf} \cdot v_{cf}^2}{2} \tag{4.57}$$

The heat created within the grinding process is removed from the contact area between grinding wheel and workpiece via the cutting fluid volume flow. Therefore, a sufficient fluid flow is necessary to remove the heat and to prevent damage to the workpiece and grinding wheel. Recommendations with regard to the necessary amount vary and are related to grinding wheel width or grinding power (for example 4 l/(min•mm) or 8–10 l/(min•kW) in continuous dressing creep feed grinding) (Webster 2007, pp. 1667f.). However, grinding width and grinding power change during the process. As a result, the supplied volume flow exceeds the needed amount if a fixed fluid volume flow is defined.

In order to match the dynamic cutting fluid demand in the contact area, the minimum volume flow rate model proposed by Metzger can be applied (Metzger 1986, pp. 134). The model relates Q_{cf} to grinding power from Eq. (4.53), orifice efficiency (η_{CFO}) as well as cutting fluid characteristics [density (ρ_{cf}), specific heat capacity ($c_{p,cf}$) and tolerated fluid temperature rise (ΔT_{cf})] (see Eq. 4.58). The orifice efficiency needs to be taken into account as not all the applied fluid flow reaches the contact area. This depends on the effective flow rate, wheel properties, jet quality, wheel speed and geometry of the grinding profile (Webster 2007, pp. 1667).

$$Q_{cf} = \frac{P_c}{c_{p,cf} \cdot \rho_{cf} \cdot \Delta T_{cf} \cdot \eta_{CFO}} \tag{4.58}$$

Equation (4.58) allows the linking of power demand of the grinding process level with power demand from the cutting fluid filtration and supply unit at the grinding system level. Therefore, a situation-oriented cutting fluid supply and power demand can be obtained.

Modelling of the overall power demand

The power demand of the grinding system (P_{GS}) can be described as the sum of the grinding machine ($P_{GM,j}$), the exhaust air filter ($P_{EAF,j}$), the cutting fluid filter ($P_{CFF,j}$) and the related power demand due to compressed air (P_{ca}).

$$P_{GS,j} = P_{GM,j} + P_{EAF,j} + P_{CFF,j} + P_{ca} \tag{4.59}$$

j stands for idle or processing (proc, d or proc, g) state.

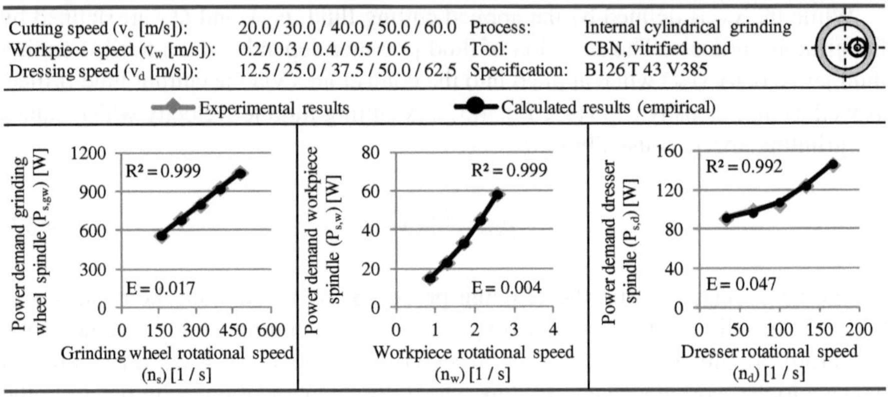

Fig. 4.26 Comparison of experimentally and empirically calculated power demand of different spindles

Verification of the physical and empirical models

In comparison with the verification step in Sects. 4.4.2.2, 4.4.2.3 and 4.4.2.4, the following verification procedure focuses on the components with a variable operational state, with a particular focus on the empirical models of the spindle power demand and the power demand of the controllable pumps.

Figure 4.26 presents the comparison between the observed and the empirically calculated power demand of a grinding wheel (left chart), workpiece (centre chart) and dresser (right chart) spindle. The values are plotted over increasing rotational spindle speeds. The observed values were obtained from an internal cylindrical grinding machine. The rotation of the grinding wheel spindle, in particular, results in a higher and nearly linear rising power demand compared to the lower and more progressive increase of dresser spindle and workpiece spindle. However, if power demand is related to the rotational speed (to determine the torque $M_i = P_i/(2 \cdot \pi \cdot n_i)$), then the workpiece spindle has the highest torque, followed by grinding and dressing spindle. The first analysis indicates a very good match between the observed and calculated results. The high R^2 and low E values confirm this result between the measured and calculated values for all spindles.

The correspondence between pump performance and empirically modelled values is presented in Fig. 4.27. The upper chart shows the mechanical pump power demand for pumping a fluid with a kinematic viscosity of 1 mm^2/s while the lower chart shows power demand for a fluid with a kinematic viscosity of 20 mm^2/s. Each chart is further divided into a drive speed of 1450 1/min on the left side and 2900 1/min on the right side. The values are plotted over an increasing fluid pressure and the consequently reduced fluid volume flow. A higher drive speed results in a higher fluid volume flow and a higher power demand compared to the lower drive speed. Furthermore, the volume flow increases with increasing fluid viscosity, and so does power demand. The regression parameters of the empirical model are presented in Appendix A.1.

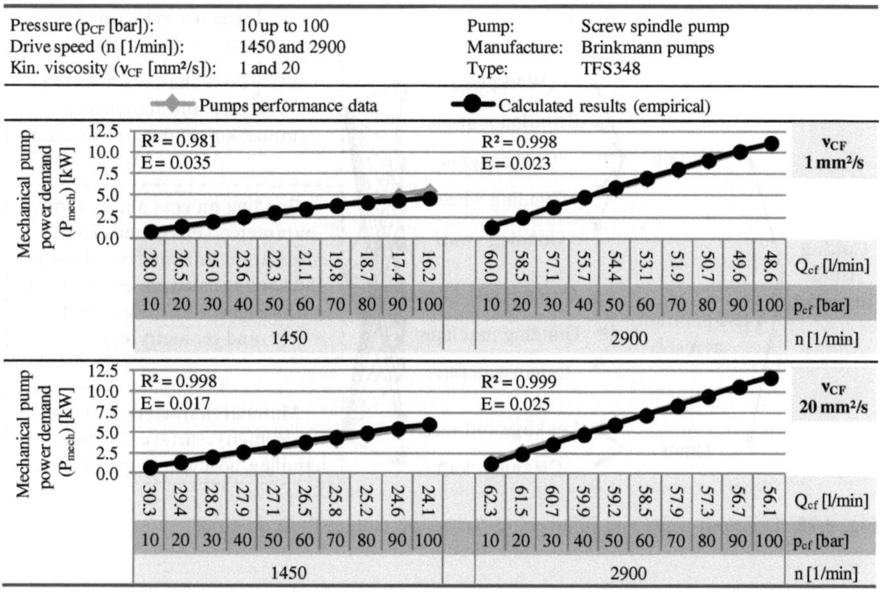

Fig. 4.27 Comparison of pump performance data and empirically calculated pump power demand

At a first glance, the calculated results indicate a very good match with the values from the pump performance data. However, with a kinematic viscosity of 1 mm^2/s and a drive speed of 1450 1/min, the chart shows a small deviation between the performance data and the empirical data at high fluid pressure. Thus, the other charts indicate a good correspondence. When considering the R^2 and E values, the first impression of matching values is confirmed. The R^2 values are high and the E values are low, indicating a good level of accuracy.

A general comparison of the physical and the empirical models is presented in Sect. 4.4.2.7.

4.4.2.6 Cutting Fluid Demand

Characteristics of the real world

The application of cutting fluid is connected with an operational state-dependent loss, also called drag-out of fluid. This drag-out results in a change of the cutting fluid's chemical and physical characteristics and influences the operating ability of the grinding system (e.g. machine down time due to insufficient cutting fluid). Accordingly, there is a need to refill the lost cutting fluid with the original cutting fluid, base fluid or additives.

Based on a literature study and on own experiments, common ways of cutting fluid loss are wetting, evaporation and aerosols as well as others (Dettmer 2006,

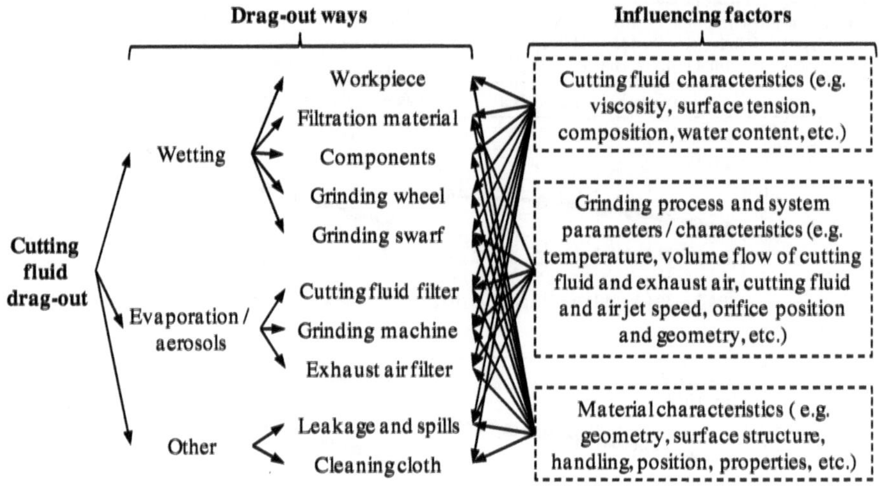

Fig. 4.28 Drag-out ways and influencing factors on cutting fluid drag-out

pp. 21; Petuelli 2002, pp. 10ff.; Mang and Dresel 2007, pp. 505). These examples for drag-out are co-presented with the influencing factors in Fig. 4.28. The wetting describes cutting fluid adhesion on workpiece, tools, components, grinding swarf and filtration material. The evaporation and aerosol generation occurs in the three considered sub-systems of the grinding system. Furthermore, cutting fluid loss can occur due to leakage of pipes, connecting elements or the tank as well as the cleaning cloth.

The drag-out itself is influenced by parameters and characteristics of the cutting fluid, the grinding process and system and the applied materials. In case of cutting fluid characteristics, fluid viscosity and surface tension are especially important. Studies showed that a high viscosity leads to high drag-out, due to better wetting of the cutting fluid on the metal surface (Petuelli 2002, pp. 27f.) and a low surface tension results in a better wetting (Kloke 2003, pp. 99). Another influence is the cutting fluid composition and water content on the evaporation behaviour of the cutting fluid.

Via the selected system parameters (volume flow as well as velocity of cutting fluid and exhaust air), the grinding process and system characteristics influence the drag-out via evaporation and aerosol generation. If cutting fluid is applied directly to the grinding wheel, for instance, a high rotational spindle speed, cutting fluid volume flow, fluid jet velocity or an unfavourable orifice position lead to a higher aerosol generation. The design characteristics also foster cutting fluid drag-out, for example via an unfavourable geometry and position of scooping elements (e.g. drilled holes, grooves) on the workpiece or grinding wheel. The handling and position of the workpiece and its components influence cutting fluid draining. The material properties and surface structure of the workpiece, grinding swarf and filtration material influence drag-out as well.

Cutting fluid pressure (p_{CF} [bar]):	ca. 6.70	Process:	Internal cylindrical grinding
Cutting fluid volume flow (Q_{CF} [l/min]):	ca. 30.00	Tool:	CBN, vitrified bond
Exhaust air volume flow (Q_{EA} [m³/min]):	11.57	Specification:	B126 T 43 V385
Material removal rate (Q'_w [mm³/s]):	10	Workpiece:	1.3505 (DIN 100Cr6)
Material removed (V_w [mm³]):	1000	Exhaust air filter:	Mechanical separation
Cutting speed (v_c [m/s]):	50	Cutting fluid:	Mineral oil emulsion / Grinding oil

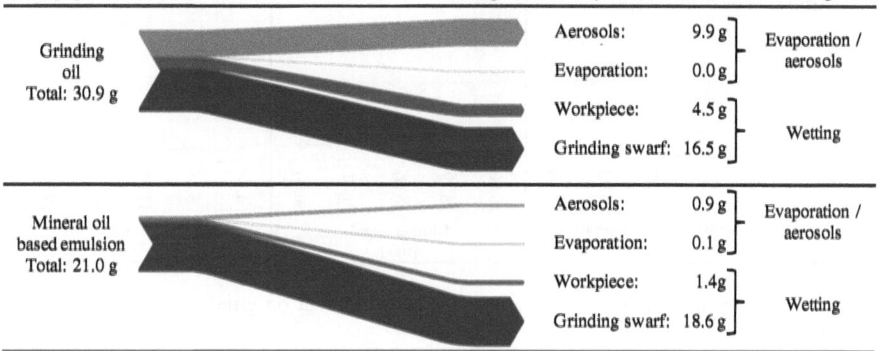

Fig. 4.29 Drag-out distribution according to cutting fluid characteristics

Figure 4.29 shows with a Sankey diagram the impact of the cutting fluid characteristics on drag-out distribution. The upper Sankey diagram shows the drag-out distribution for grinding oil and the lower Sankey diagram for a mineral oil-based emulsion. The drag-out flows are derived from an internal cylindrical grinding process of a workpiece featuring a ring shape without any scooping elements. The grinding process took 100 s, during which 7.8 g of workpiece material was removed. An exhaust air filter with a volume flow of 11.57 m³/s and cutting fluid filter with a volume flow of 30.0 l/min was connected to the grinding machine. The cutting fluid drag-outs were determined according to measurements. The drag-out via the aerosols was obtained using the procedure described by Madanchi et al. (2015, pp. 329ff.). Cutting fluid evaporation was approximated for the given process time based on long-term measurements of the cutting fluid fill level in the cutting fluid filter. Drag-out via workpiece wetting was determined by comparing the workpiece weight with a dry surface and wetted surface. A similar approach was used with the grinding swarf by comparing the dry swarf with the fluid-wetted grinding swarf.

The figure shows the influence of cutting fluid characteristics on drag-out distribution. The high viscosity ($\nu_{cf,oil} = 10.2\ \text{mm}^2/\text{s}$) of the grinding oil results in a very good wetting of the workpiece compared with the mineral oil-based emulsion ($\nu_{cf,emulsion} = 0.87\ \text{mm}^2/\text{s}$). The evaporation of both fluids was very low and the grinding oil in particular did not evaporate at all. The mineral oil-based emulsion showed a higher grinding swarf wetting compared with the grinding oil. This can be explained with the slightly higher specific surface tension of the emulsion ($\sigma_{cf,emulsion} = 31.2$ mN/m) compared to the oil ($\sigma_{cf,oil} = 29.7$ mN/m). This result is contradictory, however, because the higher viscosity of the oil should lead to higher grinding swarf wetting. It can either be assumed that surface tension has a higher

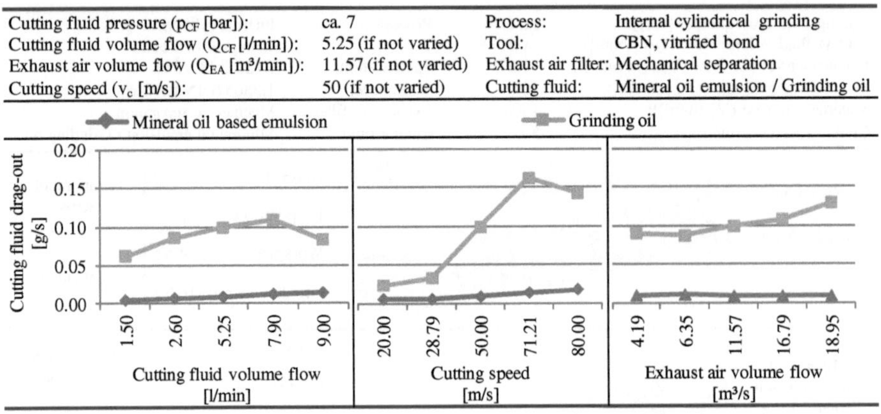

Fig. 4.30 Cutting fluid drag-out via the exhaust air dependent on grinding process and system characteristics

influence than viscosity or that this result is due to a different factor. The grinding oil showed a very high affinity with regard to aerosol formation, which is also connected to fluid viscosity and surface tension.

The characteristics of the grinding process and system are a further influence, especially on the drag-out via the exhaust air. Figure 4.30 presents the influence of the cutting fluid volume flow (left chart), the cutting speed (centre chart) and exhaust air volume flow (right chart) in dependence with two cutting fluids (grinding oil and mineral oil-based emulsion) on the exhaust air-related fluid drag-out. The presented experimental values are based on the arithmetic mean of three independent measurements.

The increase of the cutting speed in the centre chart results in rising drag-out, because the increasing grinding wheel rotational speeds result in a higher atomisation of the cutting fluid, which is then sucked away by the exhaust air filter. In case of increasing volume flows of exhaust air and cutting fluid the rate of evacuated aerosols rises as well. When comparing grinding oil and emulsion, a significant difference can be observed with regard to the drag-out rate. This difference arises due to the higher viscosity of the grinding oil. Furthermore, compared to oils with the same viscosity, the grinding oil showed a very high aerosol generation.

Abstraction of the real world into physical and empirical models

The presented characteristics of the real world showed the influence of parameters and characteristics of cutting fluid, grinding process and system as well as further materials used. Empirical as well as physical models used to predict cutting fluid demand needs to consider these influencing factors. In order to simplify the modelling process, the following assumptions apply:

- in case of a water miscible cutting fluid, only the evaporation due to water content ought to be considered,

- the influence of the grinding process parameters on chip shape is negligible when calculating the related drag-out,
- due to a leakage-free pipe and tank system, only drag-out via wetting as well as evaporation and aerosols needs to be considered
- cutting fluid drag-out via the grinding wheel is negligible. The cutting fluid is slung away from the grinding wheel in a rotation process after use.

Based on these assumptions, the overall cutting fluid demand due to loss ($\dot{m}_{cf,l}$) can be calculated based on the drag-out via the workpiece ($\dot{m}_{cf,w}$), the grinding swarf ($\dot{m}_{cf,s}$), the filtration material ($\dot{m}_{cf,fm}$), the aerosols ($\dot{m}_{cf,a}$) and the evaporation ($\dot{m}_{cf,e}$).

$$\dot{m}_{cf,l} = \dot{m}_{cf,w} + \dot{m}_{cf,s} + \dot{m}_{cf,fm} + \dot{m}_{cf,a} + \dot{m}_{cf,e} \tag{4.60}$$

The wetting of the workpiece, grinding swarf and filtration paper with cutting fluid can be modelled on the basis of Petuelli's study (Petuelli 2002, pp. 29) (see Eq. 4.61). The equation was initially proposed by Petuelli to determine the drag-out of the workpiece only. However, it can be applied to predict the drag-out via grinding swarf and filtration material as well. The formulation for drag-out due to wetting ($\dot{m}_{cf,i}$) can be described as an empirical regression model based on regression coefficients ($a_{cf,i}$, $b_{cf,i}$ and $c_{cf,i}$) and dripping time (t_{cf}).

$$\dot{m}_{cf,i} = a_{cf,i} + \frac{b_{cf,i}}{t_{cf}} + c_{cf,i} \cdot \sqrt{t_{cf}} \tag{4.61}$$

is valid for $3 < t_{cf}[s] < 3600$ with i for workpiece (w), grinding swarf (s) and filtration material (fm).

Figure 4.30 shows the significant influence of cutting fluid as well as grinding process and system characteristics influence the fluid drag-out via aerosols ($\dot{m}_{cf,a}$). To account for this influence, the nonlinear regression modelling formulation presented in Eq. (2.8) was selected. The regression model consists of cutting fluid characteristics kinematic viscosity (ν_{cf}) and surface tension (σ_{cf}) as well as process and system parameter grinding speed (v_c), volume flow of cutting fluid (Q_{cf}) and exhaust air (Q_{ea}).

$$\dot{m}_{cf,a} = K_{cf,a} + C_{cf,a} \cdot \nu_{cf}^{E_{cf,a1}} \cdot \sigma_{cf}^{E_{cf,a2}} \cdot Q_{cf}^{E_{cf,a3}} \cdot Q_{ea}^{E_{cf,a4}} \cdot v_c^{E_{cf,a5}} \tag{4.62}$$

Cutting fluid evaporation can be modelled using an approach that predicts water evaporation in swimming pools. The corresponding approach, presented in VDI 2089-1 (VDI 2089-1, 2010, pp. 15), is based on the works of Sutton, Mackay and Matsugu, as well as those of Stiver and Mackay (Sutton 1934, pp. 717; Mackay and Matsugu 1973, pp. 435; Stiver and Mackay 1984, pp. 834). The evaporation of fluids in swimming pools and in grinding systems shows three similarities: the temperature of the fluid is higher than ambient temperature, the air is in motion and the fluid surface is in motion as well. A major difference can be found in fluid composition. However, the VDI approach can still be applied when assuming that only the water content of a cutting fluid evaporates. Equation (4.63) presents the model used to calculate the evaporated cutting fluid ($\dot{m}_{\mathrm{cf,e}}$).

$$\dot{m}_{\mathrm{cf,e}} = \frac{k_{\mathrm{m}}}{R_{\mathrm{s,H_2O}} \cdot \overline{T}} \cdot \left(p_{\mathrm{s,H_2O}} - p_{\mathrm{s,air}}\right) \cdot A_{\mathrm{cf}} \cdot \theta_{\mathrm{cf}} \tag{4.63}$$

The equations include a water transfer coefficient (k_{m}), the specific gas constant for water vapour ($R_{\mathrm{s,H_2O}} = 461.52$ J/(kg K)), the arithmetic mean of cutting fluid and air temperature ($\overline{T}$), the saturation pressure of water vapour at water temperature ($p_{\mathrm{s,H_2O}}$), the water vapour pressure of the ambient air ($p_{\mathrm{s,air}}$), the surface area of evaporating cutting fluid (A_{cf}) and the water content (θ_{cf}).

The saturation pressure of water vapour at water temperature and the water vapour pressure of ambient air can be determined using the Magnus-Equation (Alduchov and Eskridge 1996, pp. 608). Equation (4.64) formulates $p_{\mathrm{s,H_2O}}$ based on the mean cutting fluid temperature ($\overline{T}_{\mathrm{cf}}$), while Eq. (4.65) describes $p_{\mathrm{s,air}}$ depending on ambient air temperature (T_{a}) [alternatively machine tool interior temperature (T_{GM})] and the relative air humidity (ϕ_{air}). Both equations are valid for temperatures between−40 °C and 50 °C.

$$p_{\mathrm{s,H_2O}} = 6109.4 \cdot e^{\frac{17.625 \cdot \overline{T}_{\mathrm{cf}}}{243.04 + \overline{T}_{\mathrm{cf}}}} \tag{4.64}$$

$$p_{\mathrm{s,air}} = 6109.4 \cdot e^{\frac{17.625 \cdot T_{\mathrm{a}}}{243.04 + T_{\mathrm{a}}}} \cdot \phi_{\mathrm{air}} \tag{4.65}$$

The VDI 2089-1 gives several values for k_{m} (also known as $\beta_{\mathrm{u/b}}$) according to the considered swimming pool, water attraction and use scenario. However, k_{m} can also be empirically calculated using the equations presented by Sutton as well as Mackay and Matsugu (Sutton 1934, pp. 717; Mackay and Matsugu 1973, pp. 435) (see Eq. 4.66).

$$k_{\mathrm{m}} = 0.0292 \cdot \mathrm{Sc}^{-0.67} \cdot v_{\mathrm{air}}^{\frac{2-n}{2+n}} \cdot d_{\mathrm{eq,cf,e}}^{\frac{-n}{2+n}} \tag{4.66}$$

The equation includes the Schmidt number (Sc), air velocity (v_{air} (in metre per hour)), the equivalent diameter of the evaporation fluid area ($d_{\mathrm{eq,cf,e}}$ (in metre)) and the exponent (n). The exponent is a function of ground roughness and temperature profile in the atmosphere (Mackay and Matsugu 1973, pp. 435). A value of $n = 0$ represents a constant air velocity at each height above the evaporation fluid. According to Mackay and Matsugu, a value of $n = 0.25$ is reasonable for average atmospheric conditions (Mackay and Matsugu 1973, pp. 435). In case of the grinding system, with an exhaust air system creating a constant velocity of sucked out air and cutting fluid which is supplied through almost the entire grinding machine, it can be assumed that $n = 0.25$ is reasonable as well. The Schmidt number represents the ratio of the kinematic viscosity of air (ν_{air}) to the diffusion coefficient (D) of the diffusing gas into air.

$$Sc = \frac{\nu_{air}}{D} \tag{4.67}$$

Verification of the physical and empirical model

The comparison between predicted and observed values for the wetting of workpiece, chips and filtration material is presented in Fig. 4.31. The two upper charts show the comparison in terms of workpiece wetting. The two centre charts present the wetting of the grinding swarf while the lower charts show the filtration material. The drag-out via workpiece, grinding swarf and filtration material wetting was calculated by comparing the weight during the dry state with the weight in the wetted state.

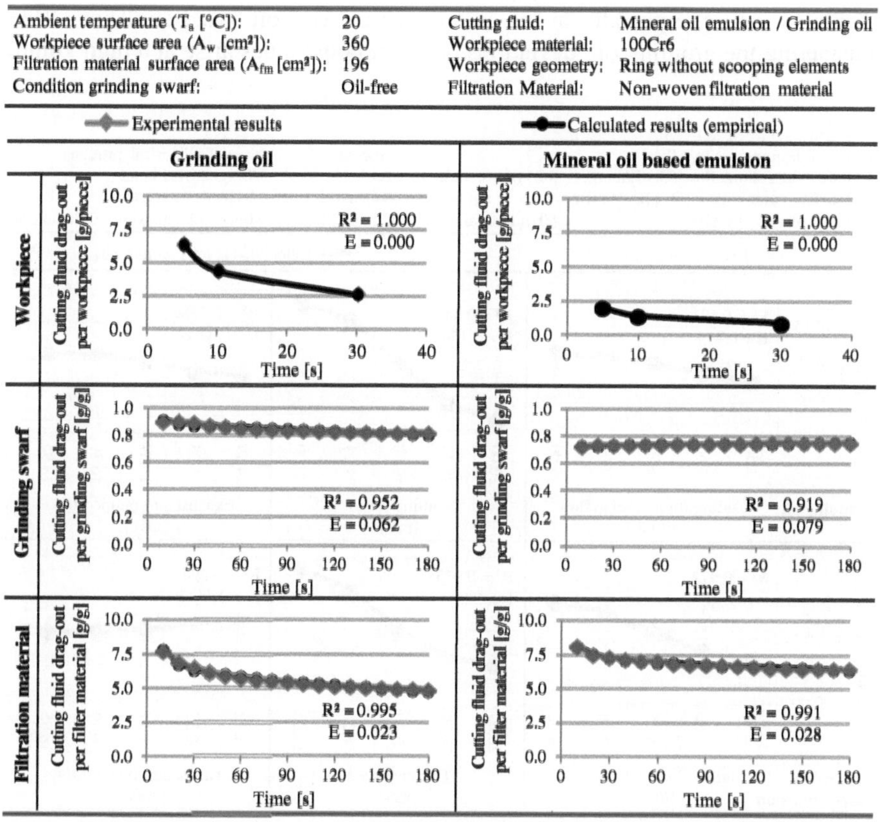

Fig. 4.31 Comparison between experimentally and empirically calculated results of the cutting fluid drag-out via wetting of surfaces

Within the first second of the experiment, the largest share of cutting fluid drips from the workpiece, grinding swarf and the filtration material. The high loss in the first seconds can be disregarded, however, because the workpiece is still in the machine tool and the filtration material (together with the grinding swarf) is in the cutting fluid filter. After 30 s, the drag-out appears to be constant on the trend. Differences are visible with regard to the influence of the cutting fluid composition. The application of the grinding oil results in a comparably higher drag-out than with emulsion in the case of the workpiece and grinding swarf, however, that is not the case for the filtration paper. The first analysis as well as the observation of the R^2 and E values indicates a good match between the observed and empirically calculated values.

Figure 4.32 shows the results of the comparison between the experimentally and empirically calculated cutting fluid drag-out via aerosols. The three upper charts show the comparison for the grinding oil and the lower three charts for the mineral oil-based emulsion. The charts show the drag-out in connection with cutting fluid volume flow (left), cutting speed (centre) and exhaust air volume flow (right).

A first analysis indicates a very good level of accuracy between the observed and calculated values, especially in the case of the mineral oil-based emulsion. When considering the grinding oil, deviations between both values arise at a high volume

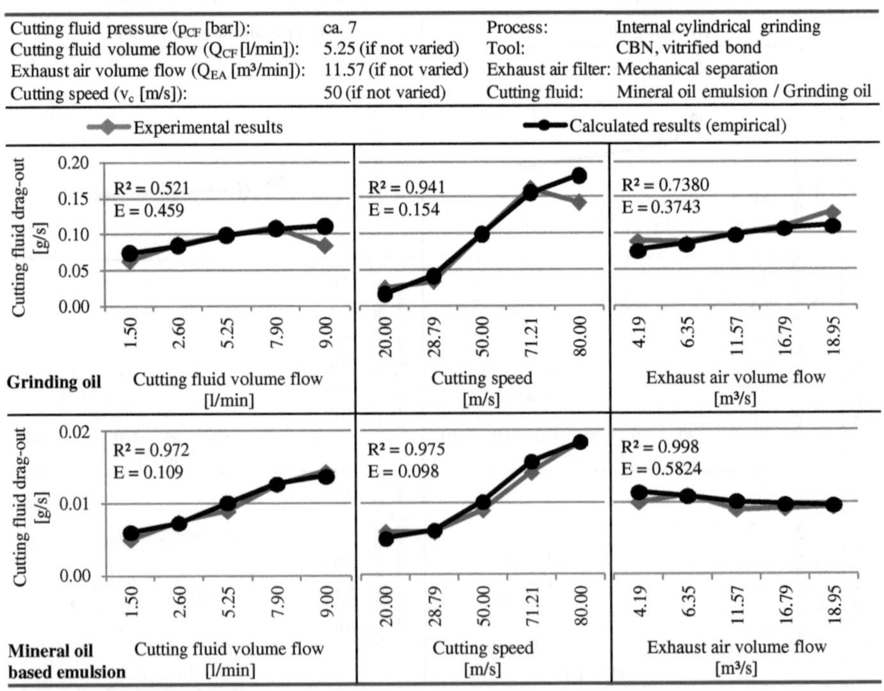

Fig. 4.32 Comparison between experimentally and empirically calculated results of the cutting fluid drag-out via aerosols

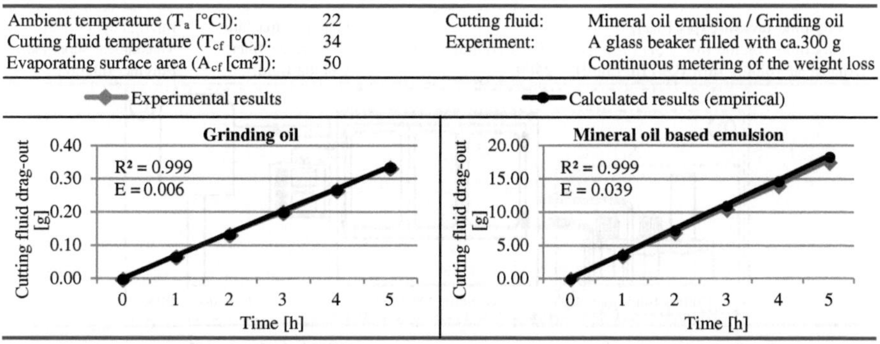

Fig. 4.33 Comparison between experimentally and empirically calculated results of the cutting fluid drag-out via evaporation

flow of exhaust air and cutting fluid as well as high cutting speeds. This could be caused by a measurement error. Each observed value, however, represents the mean value of three measurements. This deviation has an impact on the R^2 and E values, especially for the right and centre chart related to the grinding oil. Thus, if the last values were to be calculated according to the chart, then R^2 would be higher and the E value lower. In this case, the mineral oil-based emulsion indicates the high R^2 and low E values a very high accuracy.

In order to evaluate the accuracy of Eqs. (4.63) to (4.67), Fig. 4.33 shows the comparison of observed and physical calculated values to quantify the cutting fluid drag-out via evaporation. The left chart shows the comparison for the application of grinding oil and the right chart for the application of a mineral oil-based emulsion. Both charts are plotted over the time and show a linear increase. To determine the cutting fluid evaporation, a glass beaker was filled with 300 g cutting fluid and heated to 34 °C. It was assumed that $n = 0.0$, because the fluid was stored in a room with constant, very low air velocity. The evaporating fluid surface area was not in motion and the weight loss was metred over five hours.

A major difference is the significantly lower drag-out of the grinding oil compared with the mineral oil-based emulsion. When using the emulsion the water content evaporates, while in case of the grinding oil almost no water is in the oil itself. At a first glance, the calculated and observed values match very well. This impression is confirmed when considering the very high R^2 and very low E values.

This match in the values was also confirmed by other authors who applied the same equations (Fingas 2012, pp. 249ff.; Beverley 2000, pp. 1ff.). However, the conditions under which the presented results were obtained were constant. In case of the grinding process, the motion of air and cutting fluid needs to be taken into account. Therefore, Fig. 4.34 presents the impact of the grinding system on cutting fluid evaporation. Three scenarios are presented in the figure. Scenario A only focuses on the cutting fluid filter. Scenario B includes cutting fluid filter and grinding machine. Scenario C comprises the complete grinding system. The figure

Cutting fluid pressure (p_{CF} [bar]):	6.7	Grinding machine:	Internal cylindrical
Cutting fluid volume flow (Q_{CF} [l/min]):	30+30 (filtration +supply)	Exhaust air filter:	Mechanical separation
Exhaust air volume flow (Q_{EA} [m³/min]):	2500	Cutting fluid:	Polymer dilution

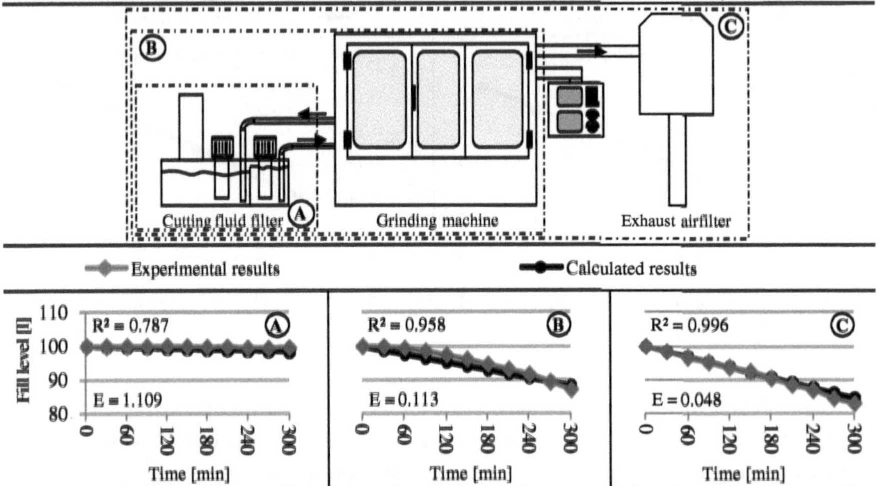

Description	Scenario	CFF $\dot{m}_{CF,e}$	GM $\dot{m}_{CF,e}$	Sum $\dot{m}_{CF,e}$	Unit
CFF	A	0.33	0.00	0.33	l/h
CFF + GM	B	0.10	2.23	2.33	l/h
CFF + GM + EAF	C	0.05	3.01	3.06	l/h

CFF: Cuttingfluid filter GM: Grinding machine EAF: Exhaust airfilter

Fig. 4.34 Influence of the grinding system on cutting fluid evaporation

shows the system borders of the three scenarios in the upper part. The centre part shows the scenario charts of the cutting fluid fill level plotted over the time, while the lower part shows the calculated cutting fluid drag-out via evaporation for each scenario as well as cutting fluid filter and grinding machine.

In scenario A, the cutting fluid is circulated within the closed filtration system only. It can be assumed that almost no air exchange occurs with the environment, the air velocity is close to zero and a quasi-steady-state condition is reached. The surface area of the evaporating fluid was 0.64 m^2. The mean drag-out loss due to evaporation is comparatively low at 0.33 l/h. The reason is to be found in a cutting fluid temperature rise and therewith increasing evaporation. The chart for the observed and calculated drag-out shows a good level of accuracy in a preliminary analysis. However, when considering the R^2 and E values, it can be concluded that the results are only partly consistent. The reason is a lower evaporation loss of the calculated evaporation loss compared with the observed one over time.

In Scenario B, the cutting fluid is circulated between the filtration system and the grinding machine and the exhaust air filter does not operate. The chart in the centre of the figure shows the comparison between the observed and calculated drag-out loss via evaporation. The observed drag-out loss changes, after an initially slow decrease (first 90 min), into a constantly declining trend. Within these 90 min, a quasi-steady-state condition is reached. As previously mentioned, it can be assumed that almost no air exchange occurs with the environment. However, due to the circulation of the cutting fluid through the grinding machine, the fluid itself is in motion. Connected with the increased surface area of evaporation (circa 1 m^2) a higher drag-out loss results compared with scenario A. The calculated drag-out loss of the cutting fluid filter is lower, as a result of decreased fluid temperature. High R^2 and low E values indicate a good match between observed and predicted values.

Scenario C considers the evaporation of the complete grinding system. The exhaust air filter evacuates the grinding machine interior with a constant air velocity of 10 m/s. Accordingly, an average atmospheric condition can be assumed. The trend of the scenario shows a constant decrease of cutting fluid fill level. Again the drag-out loss decreases due to fluid evaporation from the filtration system. However, this decrease is compensated with evaporation loss via the grinding machine and exhaust air filter. A preliminary analysis shows a good match between the observed and the predicted values. This impression is confirmed by high R^2 and low E values. A general comparison of the physical and empirical models is presented in the next section.

4.4.2.7 Comparison of Physical and Empirical Grinding Process Modelling

The previously introduced verifications for the physical and empirical models indicate a varying level of accuracy in the prediction of the observed values. In order to achieve a high level of accuracy, an appropriate modelling approach needs to be identified. Therefore, Table 4.2 presents a comparison with regard to the prediction accuracy of physical and empirical models in case of the following technological indicators: F'_n, F'_t, P'_c, R_a, V'_t, P_{GS} and $\dot{m}_{cf,l}$. The accuracy in predicting the values is assessed on the basis of the obtained R^2 and E values.

When considering the E value, the empirical models show a high level of prediction accuracy for all technological indicators; the same applies (with

Table 4.2 Comparison between prediction accuracy of physical and empirical models

	Physical model		Empirical model	
	R^2 value	E value	R^2 value	E value
Specific normal force (F'_n)	0.70	0.87	0.78	0.66
Specific tangential force (F'_t)	0.77	0.69	0.64	0.45
Arithmetical mean roughness (R_a)	0.77	1.02	0.60	0.39
Specific tool wear volume (V'_t)	0.75	12.38	0.92	0.55
Specific cutting power (P'_c)	0.77	0.69	0.64	0.45
Power demand (P_{GS})	–	–	0.99	0.02
Cutting fluid demand ($\dot{m}_{cf,l}$)	0.91	0.42	0.86	0.30

reservations) for the R^2 value as well. This is not surprising as it is the result of direct derivation of the modules from the observed values. In comparison, the application of the physical model leads to accurate results in the case of F'_t and P'_c to an average level of accuracy for the F'_n, R_a, P_{GS} as well as $\dot{m}_{cf,l}$ and for a low level of accuracy for the V'_t.

Based on this comparison, it can be concluded that not every model is suitable for predicting the investigated technological indicators. The comparison indicates that the application of empirical models is the best option. However, empirical models are based on measurements which limit their applicability in comparison with the physical models. Therefore, a combined approach is pursued. For the technological indicators F'_t and P'_c, physical models are applied. Models are used to obtain the V'_t value. For the calculation of F'_n and R_a, both models can be applied. However, in the case of F'_n the physical model is preferred over the empirical model, due to the dependence of F'_t and F'_n on the friction coefficient. As far as P_{GS} and $\dot{m}_{cf,l}$ are concerned, a combined modelling approach can be applied, based on the application of empirically and physically based equations.

4.4.3 Economic Modelling

As already mentioned in Sect. 2.2.3, three different costing approaches can be distinguished: the MFCA, the CCA and the LCC. Based on this presentation in Sect. 2.2.3, the material flow costing according to DIN EN ISO 14051 was chosen to calculate the costs of the grinding process and system. Compared with conventional cost accounting, the MFCA allows process costs to be assigned to the created product and costs to undesirable outputs. Therefore, the MFCA allows for the reduction of the share of overhead costs, due to a source-related allocation of energy and auxiliary material costs. Unlike with the CCA, residual material disposal costs are generally assigned to the overhead costs and therewith allotted to all cost units. The CCA approach does not emphasise process inefficiencies and costs of material losses resulting from a limited understanding of the overall relations of costs connected to material losses (DIN EN ISO 14051 2011, pp. 39). Furthermore, the CCA disregards that material losses also involve losses of added value and lower the overall efficiency of the production system (Kunsleben and Tschesche 2010, pp. 589).

A further option for the calculation of the costs is the application of the LCC approach instead of MFCA. The LCC approach focuses on cost accounting over all life cycle phases. The focus of the modelling approach in this investigation is the application during the utilisation phase of the grinding system. Furthermore, the LCC approach is generally used to support investment decisions by comparing alternatives based on the total cost of ownership (Schlosser 2013, pp. 59). To this purpose, costs associated with the utilisation phase are estimated based on an assumed constant mode of operation (VDI 2884 2005, pp. 14f.). Therefore, the

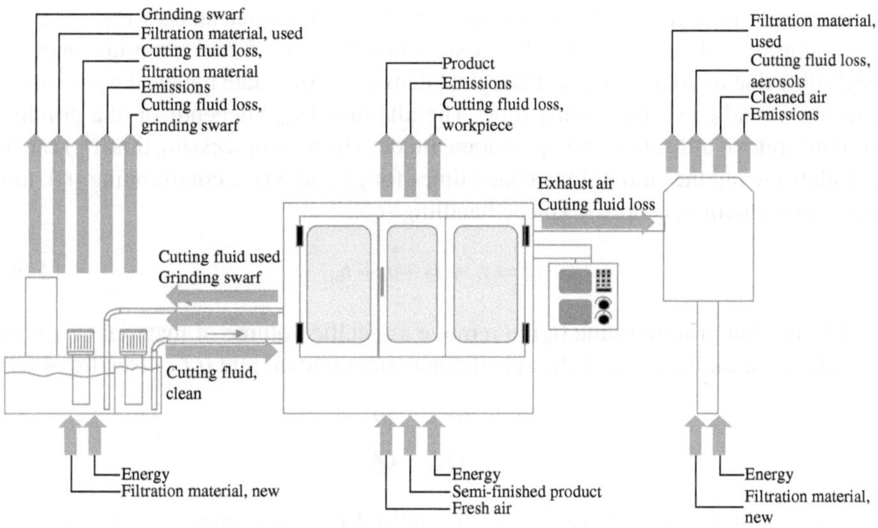

Fig. 4.35 Schematic energy and material flows of the grinding system

LCC approach is less suitable for the modelling of production processes. To this purpose, MFCA is the most appropriate approach. Should there be an interest in the life cycle, the MFCA and LCC approaches can be combined (the MFCA does not take the life cycle into account).

Characteristics of the real world

The foundation of the MFCA is the previous knowledge about material and energy flows for the investigated system. Figure 4.35 shows, in a schematic illustration comparable to Fig. 2.10, a grinding system and its related energy and material flows. The system consists of a cutting fluid filter, a grinding machine and an exhaust air filter. As already mentioned, the completion of the grinding process is connected with the transformation of energy and mass input of output flows. To apply the MFCA, it is necessary to consider the material and energetic balance of these flows. This balance is exemplarily created in Fig. 4.35 and is due to the flows of:

- energy (e.g. compressed air, electrical),
- filtration material (e.g. new, used),
- product (e.g. semi-finished, finished),
- air (e.g. fresh air, exhaust air, cleaned air),
- cutting fluid (e.g. loss, clean, used),
- emission/transmission (e.g. heat, noise).

Abstraction of the real world into physical models

For the determination of grinding system-related costs, the cutting fluid and energy demand over time need to be taken into account. As a result, it is necessary to specify the times of the operational state. In Fig. 4.25, four operational states were presented (off, ramp-up, idle, processing). The assumption that the off and ramp-up states are negligible due to no demand or a marginal demand in this state makes it necessary to consider the idle and processing time. The idle time (t_{idle}) depends on the grinding system's rate of utilisation and operational mode. The total processing time (t_p) can be calculated using the sum of the needed times for grinding (t_c), conditioning (t_d), tool travel/movement (t_t) and workpiece handling (t_h).

$$t_p = t_c + t_d + t_t + t_h \tag{4.68}$$

The needed grinding time (t_c) to remove a specific volume of material (V'_w) can be calculated on the basis of the specific material removal rate (Q'_w) (see Eq. 4.69).

$$t_c = \frac{V_w}{Q_w} = \frac{V'_w}{Q'_w} \tag{4.69}$$

The conditioning time (t_d) can be formulated in dependence with the dressing length, which corresponds, in case of a straight grinding wheel, to the tool width (b_s), the traverse dressing speed (v_{fad}) and the number of needed dressing strokes (h_d).

$$t_d = \frac{b_s}{v_{fad}} \cdot h_d \tag{4.70}$$

The number of needed dressing strokes can be determined on the basis of the radial grinding wheel wear (Δr_s) and the selected dressing depth (a_{ed}).

$$h_d = \frac{\Delta r_s}{a_{ed}} \tag{4.71}$$

The components needed to determine production costs are presented in Fig. 4.36. Based on the general structure presented in DIN EN ISO 14051, the production costs include the costs of material, energy, system and waste cost.

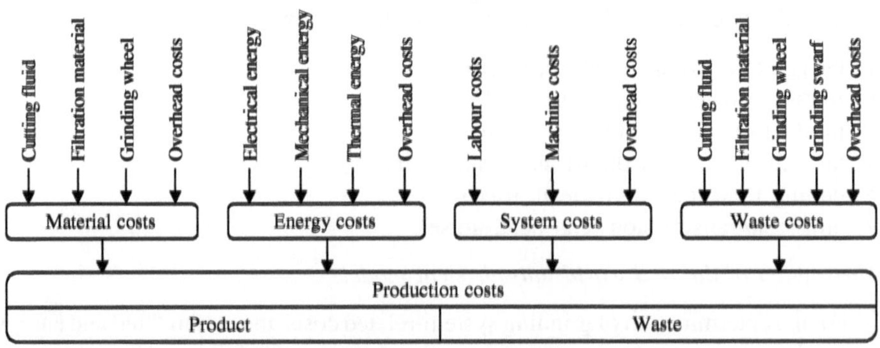

Fig. 4.36 Components used for the calculation of production costs

Material costs are calculated on the basis of procurement costs for cutting fluid, filtration material and grinding wheel as well as overhead costs (e.g. detergents, cleaning cloth). The energy costs cover the expenses for electrical, mechanical and thermal energy as well as overhead costs (e.g. heating, light). The costs for labour, machine and overhead (e.g. transportation) are merged with system costs. The waste costs include expenses for the disposal of cutting fluid, filtration material, grinding wheel residues and grinding swarf, as well as overhead costs (e.g. used detergents, used cleaning cloth).

To simplify the economic modelling, the following assumptions are made:

- the overhead costs are considered as a constant factor for each cost category,
- the costs due to emissions/transmissions (e.g. heat, noise) are included in the overhead costs for material, energy, system and waste. Thus, the requirements with regard to a levelled input–output balance are fulfilled,
- the grinding wheel abrasive and bond particles are disposed together with the grinding swarf. Therefore, the same disposal price applies for both. The disposal of the wheel hub material is considered separately.
- defective goods are not considered.

In accordance with Fig. 4.36, total production costs (C_{total}) can be formulated as the sum of material (C_{m}), energy (C_e), system (C_s) and waste costs (C_{w}).

$$C_{\mathrm{total}} = C_{\mathrm{m}} + C_{\mathrm{e}} + C_{\mathrm{s}} + C_{\mathrm{w}} \tag{4.72}$$

The material costs can be further divided into costs for the grinding wheel (C_{gw}), filtration material (C_{fm}) and the cutting fluid (C_{cf}) as well as material overhead costs ($c_{\mathrm{o,m}}$).

$$C_{\mathrm{m}} = \left(C_{\mathrm{gw}} + C_{\mathrm{fm}} + C_{\mathrm{cf}}\right) \cdot \left(1 + c_{\mathrm{o,m}}\right) \tag{4.73}$$

The costs connected with the grinding wheel (see Eq. 4.74), filtration material (see Eq. 4.75) and cutting fluid (see Eq. 4.76) can be further specified.

$$C_{\mathrm{gw}} = h_{\mathrm{d}} \cdot \frac{p_{\mathrm{p,gw}}}{h_{\mathrm{d,total}}} \tag{4.74}$$

$$C_{\mathrm{fm}} = A_{\mathrm{fm}} \cdot p_{\mathrm{p,fm}} \tag{4.75}$$

$$C_{\mathrm{cf}} = p_{\mathrm{p,cf}} \cdot m_{\mathrm{cf,l}} \tag{4.76}$$

Grinding wheel-related costs are formulated in dependence with the number of dressing strokes (h_{d}), the price of the grinding wheel ($p_{\mathrm{p,gw}}$) and the possible number of dressing times ($h_{\mathrm{d,total}}$). A similar formulation can be applied in the case of the filtration material by taking into account the surface area of the used filtration material (A_{fm}) and the price per m^2 of used filtration material ($p_{\mathrm{p,fm}}$). Cutting fluid costs are determined based on fluid price ($p_{\mathrm{p,cf}}$) and drag-out loss ($m_{\mathrm{cf,l}}$). $m_{\mathrm{cf,l}}$ can be formulated according to Eq. (4.60) considering the loss via swarf ($m_{\mathrm{cf,s}}$), workpiece ($m_{\mathrm{cf,w}}$), filtration material ($m_{\mathrm{cf,fm}}$), aerosols ($m_{\mathrm{cf,a}}$) and evaporation ($m_{\mathrm{cf,e}}$).

$$m_{\mathrm{cf,l}} = m_{\mathrm{cf,s}} + m_{\mathrm{cf,w}} + m_{\mathrm{cf,fm}} + m_{\mathrm{cf,a}} + m_{\mathrm{cf,e}} \tag{4.77}$$

The different drag-out losses can be calculated based on drag-out mass flow rates (see Eqs. 4.78 to 4.82). Considering the volume of removed workpiece material (V_{w}), workpiece density (ρ_{w}) and Eq. (4.61) to determine $m_{\mathrm{cf,s}}$, Eq. (4.61) can be used to formulate $m_{\mathrm{cf,w}}$ and $m_{\mathrm{cf,fm}}$ according to the used surface area of the filtration material (A_{fm}). The losses $m_{\mathrm{cf,a}}$ and $m_{\mathrm{cf,e}}$ can be described using processing time (t_{p}) as well as Eqs. (4.62) and (4.63).

$$m_{\mathrm{cf,s}} = V_{\mathrm{w}} \cdot \rho_{\mathrm{w}} \cdot \dot{m}_{\mathrm{cf,s}} \tag{4.78}$$

$$m_{\mathrm{cf,w}} = \dot{m}_{\mathrm{cf,w}} \tag{4.79}$$

$$m_{\mathrm{cf,fm}} = \dot{m}_{\mathrm{cf,fm}} \cdot A_{\mathrm{fm}} \tag{4.80}$$

$$m_{\mathrm{cf,a}} = \dot{m}_{\mathrm{cf,a}} \cdot t_{\mathrm{p}} \tag{4.81}$$

$$m_{\mathrm{cf,e}} = \dot{m}_{\mathrm{cf,e}} \cdot t_{\mathrm{p}} \tag{4.82}$$

Energy costs from Eq. (4.72) can be described on the basis of energy demand in the idle state ($P_{\mathrm{i,idle}} \cdot t_{\mathrm{idle}}$), in the grinding processing state ($P_{\mathrm{GM,proc,g}} \cdot t_{\mathrm{c}}$) and in the conditioning processing state ($P_{\mathrm{GM,proc,d}} \cdot t_{\mathrm{d}}$). Energy demand is then multiplied by the energy price (p_{e}). Furthermore, energy-related overhead costs are also taken into account ($c_{\mathrm{o,e}}$).

$$C_{\mathrm{e}} = \left[p_{\mathrm{e}} \cdot \left(P_{i,\mathrm{idle}} \cdot t_{\mathrm{idle}} + P_{\mathrm{GM,proc,g}} \cdot t_{\mathrm{c}} + P_{\mathrm{GM,proc,d}} \cdot t_{\mathrm{d}}\right)\right] \cdot \left(1 + c_{\mathrm{o,e}}\right) \tag{4.83}$$

To calculate system-related costs, the following elements are considered: expenses for the grinding system (p_{m}) and labour (p_{l}) as well as system overhead costs ($c_{\mathrm{o,s}}$). p_{m} can be calculated using the machine-hour rate based on imputed depreciation, imputed interest, occupancy costs and maintenance costs, presented for example in Plinke and Rese (2006, pp. 129). Furthermore, p_{m} represents the sum of the machine-hour rate of grinding machine ($p_{\mathrm{m,GM}}$), cutting fluid filter ($p_{\mathrm{m,CFF}}$) and exhaust air filter ($p_{\mathrm{m,EAF}}$).

$$C_{\mathrm{s}} = \left[(p_{\mathrm{m}} + p_{\mathrm{l}}) \cdot t_{\mathrm{p}}\right] \cdot \left(1 + c_{\mathrm{o,s}}\right) \tag{4.84}$$

The cost of waste can be determined on the basis of the costs of disposal ($p_{\mathrm{w,cf}}$) for the used cutting fluid ($m_{\mathrm{w,cf}}$) according to the processing time (t_{p}) and cutting fluid service life ($t_{\mathrm{cf,l}}$). Furthermore, the equation includes the costs of disposal

($p_{w,s}$) for the grinding swarf, represented via the volume of removed workpiece material (V_w) and workpiece density (ρ_w), as well as the costs for removed grinding wheel debris, represented via the specific tool wear volume (V'_t), tool width (b_s) and tool density (ρ_{gw}). The disposal-related costs or revenues of the wheel hub material ($p_{w,gm}$) are considered in proportion to the grinding wheel's total number of possible dressing times ($h_{d,total}$) and the actual dressing times (h_d).

$$C_w = \left[p_{w,s} \cdot \left(V_w \cdot \rho_w + V'_t \cdot b_s \cdot \rho_{gw}\right) + \frac{p_{w,gm}}{h_{d,total}} \cdot h_d + p_{w,fm} \cdot A_{fm} + p_{w,cf} \cdot m_{w,cf} \cdot \frac{t_p}{t_{cf,l}}\right] \cdot \left(1 + c_{o,w}\right) \tag{4.85}$$

4.4.4 *Environmental Modelling*

This chapter is structured into two main sections and focuses on the environmental modelling of the grinding process and system. The first section, including the Sect. 4.4.4.1, introduces the equations used to model the environmental impact of the grinding process and system in connection with the demand of energy and material flows. Owing to the dependence of the equations on the created environmental impact, the second section presents simplified material and process flows during the life phases of grinding wheel (Sect. 4.4.4.2), cutting fluid (Sect. 4.4.4.3) and filtration material (Sect. 4.4.4.4). Based on these flows and the calculation of the environmental impact, the life cycle assessment approach based on ISO 14040 can be applied (see Sect. 2.2.3). In case of the production and distribution of electrical energy, a value from the databases (e.g. Ecoinvent, GaBi) can be used to determine the environmental impact. This also applies for grinding swarf disposal, which is usually incinerated due to its classification as hazardous waste (Dettmer 2006, pp. 27, pp. 126).

4.4.4.1 Environmental Modelling of the Energy and Resource Flows

Characteristics of the real world

Environmental modelling can be conducted in a similar manner to the economic modelling of energy and resource flows. Instead of applying financial values to measure the economic impact, the environmental impact can be used to calculate the environmental impact. The environmental impact is also connected to the energy and resource flows entering and leaving the grinding process and system. Different impact categories can be used within the impact assessment when measuring the environmental impact of energy and material flow. Corresponding impact categories are for example global warming potential, abiotic resource depletion,

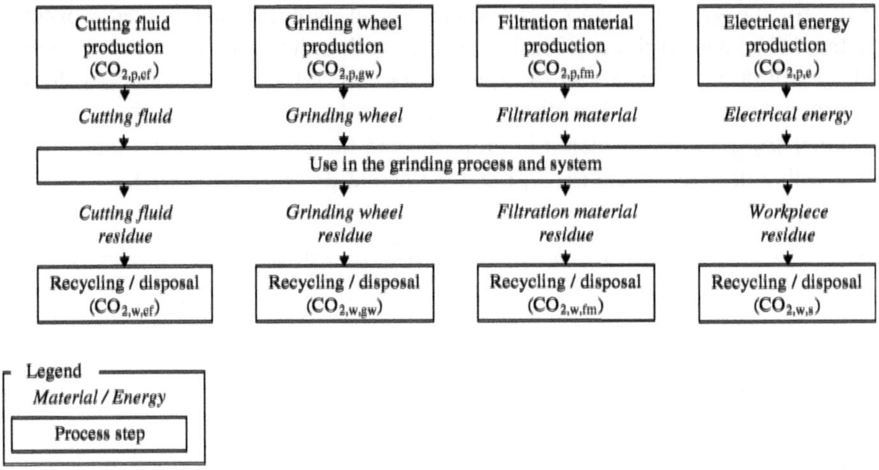

Fig. 4.37 Simplified material and process input and output flows of the grinding process and system

human toxicity and ecotoxicity (Guinée et al. 2004, pp. 529ff.). In this investigation, the global warming potential (GWP) will be used to represent the environmental impact resulting from the demand of energy and resource flows. The GWP allows the assessment of the potential contribution of a substance on climate change, but it merely provides a rough indication of the potential climatic effects of such emissions due to further influences (Guinée et al. 2004, pp. 572). However, the application of the GWP is a viable and an often used option (Thiede 2012, pp. 123). The reason is to be found in the worldwide accepted definition and quantification method of the GWP, its extensive appraisals by international peer reviews as well as the scientific measurement of corresponding equivalence factors (e.g. methane, nitrous oxide) (Klöpfer and Grahl 2009, pp. 252). The GWP or carbon footprint can be represented in a unit of kilogram of carbon dioxide equivalent (kg CO_2eq).

Accordingly, every resource and energy flow entering and leaving the grinding system contributes to the environmental impact. This impact can be expressed in a unit of kg CO_2eq. This relationship is presented in Fig. 4.37. The figure shows the equivalent CO_2 impact due to the production of cutting fluid ($CO_{2,p,cf}$), grinding wheel ($CO_{2,p,gw}$), filtration material ($CO_{2,p,fm}$) and electrical energy ($CO_{2,p,e}$). Furthermore, the impact is connected with the residue recycling or disposal of the cutting fluid ($CO_{2,w,cf}$), grinding wheel ($CO_{2,w,gw}$), filtration material ($CO_{2,w,fm}$) and grinding swarf ($CO_{2,w,s}$).

Abstraction of the real world into physical models

In the following, environmental modelling is based on the equivalent carbon dioxide impact presented above. In order to simplify environmental modelling, the following assumptions are made:

- the environmental impact due to workpiece production can be neglected,
- the environmental impact of the production, use and disposal of the grinding machine and the filtration units can be neglected,
- the grinding wheel abrasive and bond particles are disposed together with the grinding swarf. Thus, the same disposal specific equivalent CO_2 impact applies. The disposal of wheel hub material can be considered separately,
- defective goods are not considered.

The total environmental impact of the grinding process and system (Env_{total}) can be modelled, as shown earlier in Fig. 4.6, by combining the environmental impact resulting from the grinding wheel (Env_{gw}), the filtration material (Env_{fm}), the cutting fluid (Env_{cf}), the electrical energy (Env_{e}) and the waste (Env_{w}).

$$\mathrm{Env}_{total} = \mathrm{Env}_{gw} + \mathrm{Env}_{fm} + \mathrm{Env}_{cf} + \mathrm{Env}_{e} + \mathrm{Env}_{w} \tag{4.86}$$

The environmental impact of the grinding wheel (see Eq. 4.87), filtration material (see Eq. 4.88) and cutting fluid (see Eq. 4.89) is calculated in a similar way to the economic modelling presented in Sect. 4.4.2. Equation (4.87) is based on the number of dressing strokes (h_d) in relation to the equivalent CO_2 impact of grinding wheel production ($CO_{2,p,gw}$) and the possible number of dressing times ($h_{d,total}$). The Env_{fm} is formulated according to the surface area of the used filtration material (A_{fm}) and the equivalent CO_2 impact of filtration material production ($CO_{2,p,fm}$) per m^2 of used filtration material. The cutting fluid's environmental impact depends on the equivalent CO_2 impact of cutting fluid production per kg fluid ($CO_{2,p,cf}$) multiplied by drag-out loss ($m_{cf,l}$).

$$\mathrm{Env}_{gw} = h_d \cdot \frac{CO_{2,p,gw}}{h_{d,total}} \tag{4.87}$$

$$\mathrm{Env}_{fm} = A_{fm} \cdot \frac{CO_{2,p,fm}}{A_{fm,total}} \tag{4.88}$$

$$\mathrm{Env}_{cf} = CO_{2,p,cf} \cdot m_{cf,l} \tag{4.89}$$

By considering the energy demand in the idle state ($P_{i,idle} \cdot t_{idle}$), in the grinding processing state ($P_{GM,proc,g} \cdot t_c$) and in the conditioning processing state ($P_{GM,proc,d} \cdot t_d$), the Env_e can be measured depending on the equivalent CO_2 impact of electrical energy production per kWh.

$$\mathrm{Env}_{e} = \left(P_{i,idle} \cdot t_{idle} + P_{GM,proc,g} \cdot t_c + P_{GM,proc,d} \cdot t_d\right) \cdot CO_{2,p,e} \tag{4.90}$$

Env_w is formulated on the basis of the equivalent CO_2 disposal impact of cutting fluid ($CO_{2,w,cf}$), grinding wheel hub material ($CO_{2,w,gw}$), filtration material ($CO_{2,w,fm}$) and grinding swarf ($CO_{2,w,s}$). The fluid's impact is calculated in relation to the mass of disposed fluid ($m_{w,cf}$). The grinding wheel's total number of possible dressing times ($h_{d,total}$) is used to measure the wheel hub's disposal impact. The filtration material impact is formulated in connection with the used surface area of filtration material (A_{fm}). The volume of removed workpiece material (V_w), workpiece density (ρ_w) as well as the removed grinding wheel debris, represented via tool wear volume (V_t), tool width (b_s) and tool density (ρ_{gw}) are all taken into account when measuring the workpiece swarf impact.

$$\mathrm{Env_w} = \mathrm{CO}_{2,\mathrm{w,s}} \cdot \left(V_\mathrm{w} \cdot \rho_\mathrm{w} + V'_\mathrm{t} \cdot b_\mathrm{s} \cdot \rho_\mathrm{gw}\right) + \mathrm{CO}_{2,\mathrm{w,cf}} \cdot m_\mathrm{w,cf} \cdot \frac{t_\mathrm{p}}{t_\mathrm{cf,l}} + \mathrm{CO}_{2,\mathrm{w,fm}} \cdot A_\mathrm{fm} + \frac{\mathrm{CO}_{2,\mathrm{w,gw}}}{h_\mathrm{d,total}} \cdot h_\mathrm{d} \tag{4.91}$$

4.4.4.2 Grinding Wheel

Characteristics of the real world

The life phases connected with the extraction of raw materials and the production, usage and disposal of a grinding wheel have an impact on the environment due to resource and energy transformation. This impact can be considered as being part of the environmental impact of the grinding wheel. The application of the grinding wheel in the usage phase results in a virtual transfer of the aforementioned impact from the grinding wheel to the processed workpiece. This transfer can be characterised on the basis of abrasive material wear. As mentioned in Sect. 4.4.2.4, the wear rate of the grinding wheel depends on the selected process and conditioning parameters of the grinding process.

In order to determine the environmental impact of the grinding wheel, it is necessary to take into account the material and energy flows during the life phases. Studies on this topic were either theoretical (Linke and Overcash 2012, pp. 293ff.; Linke 2014, pp. 258ff.), considered the impact of specific components (e.g. abrasive grains) (Li 2012, pp. 86ff.), determined the overall grinding wheel impact of specific grinding wheel types (e.g. cBN metal bonded on a steel wheel hub) (Murray et al. 2012, pp. 1604ff.) or focused on cBN vitrified bonded wheels with varying hub materials (Winter et al. 2015, pp. 1ff.). The corresponding modelling approach for measuring the environmental impact of grinding wheels needs to consider varying combinations of abrasive grain materials, additives, bond types and wheel hub materials.

Abstraction of the real world into physical models

The exploration of raw materials and their processing into abrasive grains and wheel hub material is connected with a negative environmental impact. The main source of environmental impact is the demand of processing energy (Linke 2014, pp. 259) and the connected carbon dioxide emissions during the production phase.

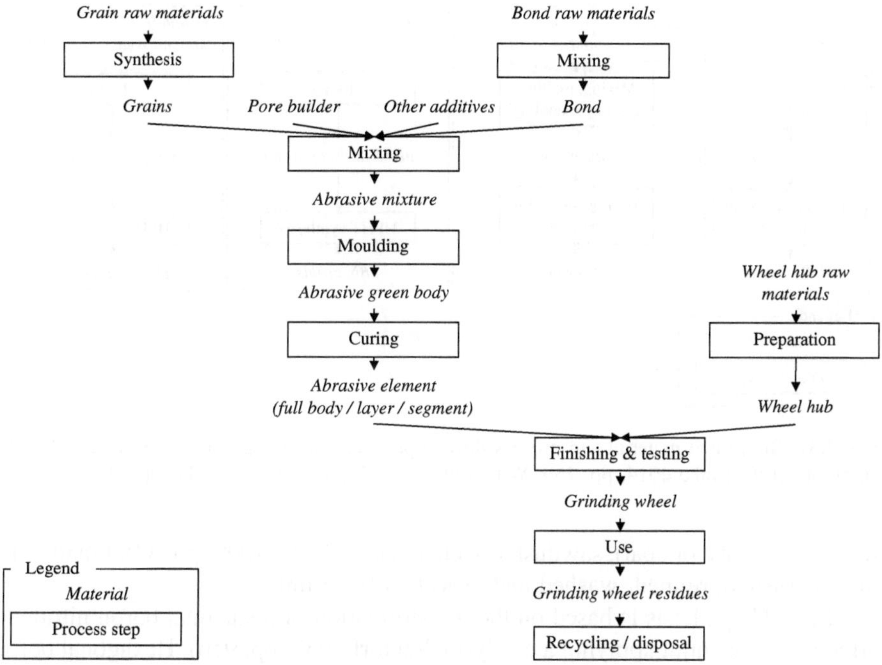

Fig. 4.38 Simplified material and process flow to produce grinding wheels

In the usage phase, the grinding wheel can influence the process' energy demand, irrespective of its design.

Figure 4.38 presents a simplified process and material flow chart for the production of a grinding wheel composed of abrasive elements and wheel hub. Furthermore, the figure includes the use and disposal phases. The abrasive elements consist of three main components, namely abrasive grains, bond and additives (pore builder and other additives). As mentioned in Sect. 2.1.2, four grain types can usually be distinguished: aluminium oxide, silicon carbide, cubic boron nitride and diamond. The grain raw materials are mixed and produced using a synthesis step. A simplified material and process flow to produce the abrasive grains is presented in Fig. 4.39.

The synthesis of aluminium oxide can be conducted by melting alumina in electric furnaces at temperatures of 2000 °C according to the process invented by Jacobs. The purity degree of alumina influences the created type of aluminium oxide and grain properties (Klocke 2009, pp. 20f.). After a processing time of 3–24 h (depending on furnace type and size), the molten aluminium oxide cools down for several days, followed by milling, cleaning and sieving. Black and green silicon carbide can be produced by melting quartz sand in a resistance furnace with the Acheson process (Linke 2014, pp. 259). The melting process has a duration of about 36 h, at temperatures of 2000–2400 °C, and requires the presence of a carbon

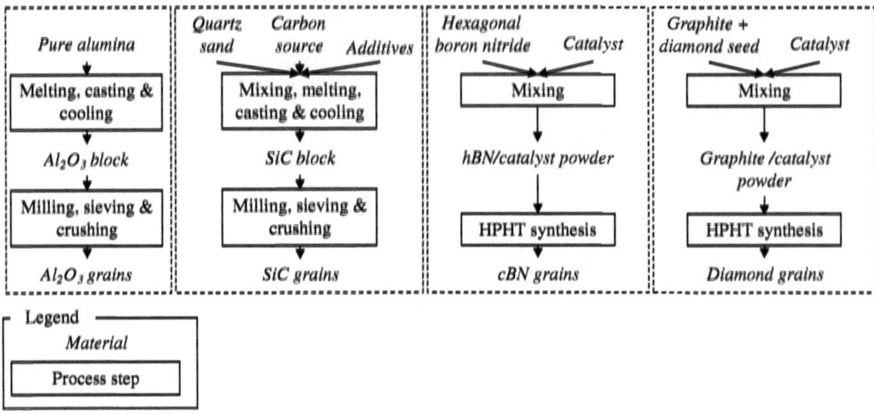

Fig. 4.39 Simplified material and process flow to produce abrasive grains (According to Klocke 2009, pp. 20ff.; Linke 2014, pp. 259; Winter et al. 2015, pp. 12ff.; Ali 2011, pp. 3)

source (e.g. coke or coal), sawdust as well as salt (Klocke 2009, pp. 33). Finally, the SiC is crushed, refined, washed and classified by sifting.

The cBN synthesis is based on the transformation of hexagonal boron nitride to cBN with or without applying a catalyst (Wentorf 1957, pp. 956). Hexagonal boron nitride (hBN) is a binary compound of boron and nitrogen featuring a layered structure with stacked sheets of six-membered rings. The transformation requires the application of high pressure and high temperature (HPHT) and takes about 15–20 min at a temperature of 1200–2000 °C and pressures of 2.5–7.5 GPa (Winter et al. 2015, pp. 12). Artificial diamonds can also be produced by using the HPHT process. In this process, small seed diamonds are covered with a mixture of catalyst and graphite. The mixture is then pressed and heated at similar pressures and temperatures as the cBN synthesis (Ali 2011, pp. 3). The aforementioned catalysts have the function of a flux precursor to create a eutectic in which the boron nitride or graphite is partially dissolved (Vel et al. 1991, pp. 155ff.). This allows the influencing of equilibrium pressure and temperature. A literature study of Guo et al. reported about 50 kinds of catalysts to support cBN synthesis (state 2010) (Guo et al. 2010, pp. 1296). The reported catalysts are based on alkali earth metals and alkali metals. Advantages of applying the catalysts are to be found in the increase of productivity and in the reduction of production costs, due to energy demand and die tool wear (Winter et al. 2015, pp. 9).

Four types of bond systems can be distinguished: vitreous bond, organic/synthetic resin bond, metallic sintered bond and electroplated bond. The vitreous bond is the most frequently applied. This bond system consists of a borosilicate glass, which is composed of aluminium oxide, boric oxide, silica, alkaline earth metals (e.g. calcium oxide, magnesium oxide) and alkaline metals (e.g. sodium oxide, potassium oxide). Plastic, polyimide resin, phenolic resin or phenol-aralkyl formulations can be used to produce the organic/synthetic resin bond systems. Bronze made of copper–tin alloys is used to produce metallic sintered bond systems. In case of the electroplated bond system, a single layer of abrasive grains is directly bonded to a metallic wheel hub by an electrodeposition of nickel or a nickel alloy.

Additives are needed to enhance or to prevent specific characteristics of the abrasive elements. One major function of the additives is to enhance the creation of pores if the amount of naturally formed pores is insufficient. The pore building additives allow the achievement of a defined porosity and permeability which is important for the metal removal process performance. Three different types of pore inducing additives/processes can be distinguished: the burn-out method, the closed cell or bubble method and the wash- or melt-out method (Winter et al. 2015, pp. 10f.). Further additives, in addition to the pore inducing ones, are used to influence specific characteristics during the manufacturing process or during the application of the abrasive elements. For example, to enhance the abrasive green body's cohesion, a temporary binder is added (e.g. dextrin, water, paraffin wax). These additives are usually thermally decomposed during the curing process. In order to keep a defined ratio of bond and abrasive grains while reducing the application of expensive primary abrasive grains (e.g. cBN or diamond), secondary abrasive grains and secondary active and inactive fillers are used (Winter et al. 2015, pp. 10f.). To this end, secondary grains made of for example garnet, flint, aluminium oxide or silicon carbide are used. Different halogen salts or metal sulphides can be applied as secondary fillers. To enhance abrasive element performance during its application, graphite, hBN or molybdenum disulphide can be added to foster a self-lubricating effect (Winter et al. 2015, pp. 10f.).

As presented in Fig. 4.38, abrasive grains, pore builders, other additives and bond are mixed in a defined ratio. This abrasive mixture is then pressed through a moulding process into an abrasive green body. The green body is then cured in the next step using either a cold pressing or hot pressing process (Winter et al. 2015, pp. 8). After the curing process, the abrasive element is mounted on the grinding wheel hub. The applied hub material varies depending on the grinding wheel design, abrasive type, size and application scenario. Usually wheel hubs made of low-alloyed steel, aluminium, bronze, ceramic or synthetic resin (fibre-reinforced; with metallic or non-metallic fillers) are applied (Klocke 2009, pp. 50f.). The last step of the process enfolds finishing and testing, and the geometry is checked with regard to hardness and unbalance as well as burst resistance and corrected if necessary (Winterthur Technology Group 2011, pp. 9).

During the grinding wheel usage phase, the grinding wheel influences grain wear, cutting forces and power as well as workpiece surface roughness, depending on the process and conditioning parameters. The interdependencies of grinding wheel and application scenarios can be taken into consideration based on either experimental measurements or based on the wear, force, power and surface roughness models presented in Sect. 4.4.2.

During the usage phase, grinding wheel wear residue is commonly disposed of together with the grinding swarf. After the usage phase, there are different options with regard to recycling or disposal of the grinding wheel. The corresponding option depends on grinding wheel design as well as abrasive and wheel hub material. A full-body grinding wheel consisting of conventional abrasive can be directly disposed of with municipal waste. Grinding wheels with a doughnut or a single abrasive layer can be disposed of (e.g. municipal waste, metal scrap) as well or sent back to the grinding wheel manufacturing for relayering. The latter occurs in the case of grinding wheels consisting of superabrasives and metal wheel hubs.

4.4.4.3 Cutting Fluid

Characteristics of the real world

The physical modelling of the cutting fluid's environmental impact is based on the comprehensive work of Dettmer (2006, pp. 121ff.). In the context of a comparative life cycle assessment, Dettmer analysed the impact of non-water miscible cutting fluids for grinding processes based on mineral oil, castor oil, palm oil, animal fat and used cooking oil.

Abstraction of the real world into physical models

A simplified material and process flow for cutting fluids is presented in Fig. 4.40. The starting point of cutting fluid production is the preparation of base fluid and additives from raw materials. In the following processing step, base fluid, additives and water (if required) are mixed. The usage and recycling or disposal of cutting fluid residues follows.

As shown in Fig. 2.7, different base fluid raw materials can be used to formulate cutting fluids. Each base fluid raw material needs different processing steps to be transformed into a base fluid. These processing steps are exemplarily presented for five different base fluid materials in Fig. 4.41. The figure shows the simplified material and process flows of base fluids, produced from crude mineral oil, vegetable oil, used cooking oil, animal fat and raw glycerol. While mineral base oil is

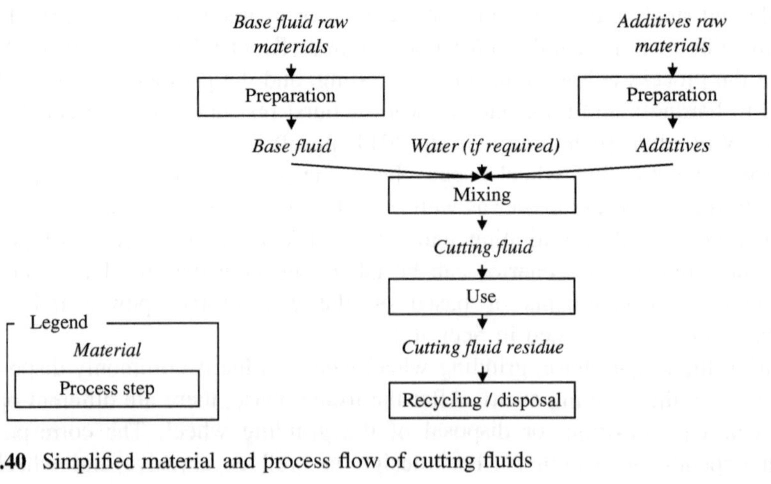

Fig. 4.40 Simplified material and process flow of cutting fluids

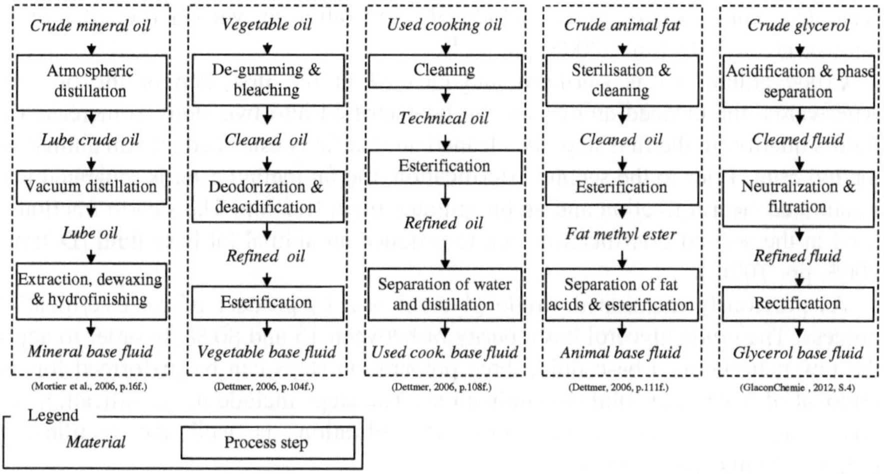

Fig. 4.41 Simplified material and process flow to produce base fluids (According to Mortier et al. 2006, pp. 16f.; Dettmer 2006, pp. 104ff.; GlaconChemie 2012, pp. 4)

the most common basic ingredient for conventional cutting fluids, the other four base fluids are a viable option for the production of alternative cutting fluids from a renewable raw material.

The production of a mineral base oil starts with the atmospheric distillation of crude mineral oil into fractions (e.g. gases, gasoline, fuels, lube oils, bitumen) at temperatures ranging from 350 to 370 °C and a pressure of 1 bar (Umweltbundesamt 2000, pp. 48ff.). The lube crude oils are then transferred into a vacuum distillation unit for a further separation of gas oil and vacuum residues from the lube oil (Mortier et al. 2006, pp. 16f.). The process is realised in a vacuum distillation unit with a pressure of 0.17 bar and 380°C at the column bottom and 0.08 bar and 65 °C at the top (Mortier et al. 2006, pp. 19). In the refinement step aromatic extracts, wax and light fuels are separated using extraction, dewaxing and hydrofinishing processes (Mortier et al. 2006, pp. 21ff.; Umweltbundesamt 2000, pp. 51ff.).

Vegetable oil (e.g. palm oil, rape seed oil, jatropha oil) can be used to produce vegetable base oils. The transformation steps from raw oil to base oil include degumming and bleaching, deodorisation and deacidification as well as esterification process (Dettmer 2006, S103ff.). The degumming and bleaching as well as deodorisation and deacidification process steps are necessary to increase fluid quality and purity (Kent 2007, pp. 1606). During the esterification step, a separation of fat acids and glycerol under high pressures and temperatures is conducted. After the separation, the saturated fat acids are esterified with the applied alcohols (e.g. 2-ethylhexanol) (Dettmer 2006, pp. 104).

The cleaning step for used cooking oil entails the separation of water and solid contaminations (Dettmer 2006, pp. 112). The created fluid has a technical grade and is used for the esterification. During esterification, glycerols and fat acids are separated. The processing step includes the separation of water and distillation to remove alcohol (Dettmer 2006, pp. 113).

Crude animal fat is sterilised and cleaned in the first step of the process. Afterwards, the cleaned animal fat can be esterified into two steps to increase the fluid's quality. In the first step, the cleaned animal fat is esterified into an animal fat methyl ester. Prior to the second esterification, the fat methyl ester is separated into a saturated stearin fraction and an unsaturated olein fraction. The stearin fraction is used in the second esterification step to produce an animal fat base fluid (Dettmer 2006, pp. 107ff.).

As previously mentioned, crude glycerol is a by-product of the esterification process. The crude glycerol has a purity of between 15 and 80 %. In order to apply the raw material as a base fluid, three preparation steps can be performed for the removal of water and solid contaminations. The steps include the acidification and phase separation, the neutralisation and filtration as well the rectification (GlaconChemie 2012, pp. 4).

While base fluid production processes are well documented in the literature, only limited information are available for the additives. Dettmer reports that the additivation of cutting fluids in particular is kept undisclosed and only limited inventory information about additives is available (Dettmer 2006, pp. 156). However, based on the overview of commonly used additives in cutting fluids presented in Baumann and Herberg-Liedtke (1996, pp. 250ff.), the material and resource flows during additive production can be approximated in connection with a patent and literature research.

As shown in Fig. 4.40, base fluid, additives and (if required) water are mixed. This step occurs under continuous stirring at varying cutting fluid temperatures in a mixing vessel to create a stable chemical compound. Accordingly, electrical energy for the stirrer and for the heating device needs to be taken into consideration (Hansen et al. 2005, pp. 19).

The duration of the usage phase depends on fluid composition, service condition, fluid renewal rate and fluid maintenance (Mang and Dresel 2007, pp. 488ff.). The cutting fluid maintenance prolongs the service life thanks to the application of technological (e.g. cleaning, filtration) and chemical (e.g. additives) measures. This is connected with energy and material demand as well. Additionally, the cutting fluid application demands energy and materials. The energy demand of cutting fluid filter and exhaust air filter can be modelled in accordance with the selected grinding process and conditioning parameters as well as grinding system parameters (see Eqs. 4.47 to 4.50). Material demand arising from cutting fluid loss can be measured using Eq. (4.60).

Whether a cutting fluid is disposed of or recycled depends on its base fluid type. Mineral oil-based fluids can either be rerefined to a new base fluid or thermally recycled. The fluids need to be cleaned and the water content removed. If rerefining is not an option, then the fluid is thermally recycled (Dettmer 2006, pp. 126). As far as water miscible fluid is concerned, the water content is separated from a mixture of base fluid, additives and tramp oils using for example evaporators or demulsification systems. The remaining compound is then thermally recycled.

Results from Dettmer's study indicate that in case of the application of mineral oil-based cutting fluids 50 % of the GWP is caused by the fluid's disposal. In case of alternative cutting fluids with a vegetable or animal base almost no GWP is caused due to CO_2 neutral thermal recycling (Dettmer 2006, pp. 135).

4.4.4.4 Filtration Material

Characteristics of the real world

No specific studies exist in literature for the environmental modelling of cutting fluid filtration material. However, extensive studies exist with regard to the impact assessment of the life phases of common paper and textiles. Due to a high level of similarity in the production processes for paper and textiles as well as filtration material, a high degree of transferability can be assumed. The following section explores the production processes of filtration material in accordance with the processes for paper and textiles in the modelling of environmental impact.

Abstraction of the real world into physical models

A simplified material and process flow to produce woven, non-woven and paper filtration material is presented in Fig. 4.42. The production steps to produce woven filtration materials encompass the spinning of the fibre raw material to yarn, followed by the weaving or knitting of the yarn to the fabric and finally a finishing step (Purchas and Sutherland 2002, pp. 35ff.). Natural (e.g. cotton, flax, silk, wool) or artificial (e.g. thermoplastic polymers, glass, ceramic, carbon) fibre raw materials can be used. The yarn can be woven into a plain, twill or satin weave fabric. To ensure fabric stability, to modify the surface characteristics, to regulate the fabric permeability and to generate the required geometry, a finishing step is conducted (Purchas and Sutherland 2002, pp. 63f.).

For the production of non-woven filtration material, man-made fibres (e.g. polypropylene, polyester, polyamide) or semi-synthetic fibres (e.g. viscose) can be used. These raw materials are applied in a melt spinning process at 280 °C to produce fibres (Fuchs and Albrecht 2012, pp. 33ff.). For the production of one kilogram of polyester fibre, for example, energy demand varies between 104 and 127 mega joule, polypropylene production requires about 70–80 MJ/kg and polyamide production around 120–138 MJ/kg (Muthu 2014, pp. 14ff.). The fibres can then, either directly or indirectly, be transferred to the web formation process to produce unbounded fibres. Web formation can be achieved using the drylaid, spunlaid and the wetlaid processes (Fuchs and Albrecht 2012, pp. 119ff.). To increase the mechanical properties of non-woven material, a web bonding process step can be conducted. Chemical (e.g. acrylate binders), mechanical (e.g. needle punching, hydroentanglement) or thermal (e.g. calendering process, hot air) bonding processes can be used in this step (Fuchs and Albrecht 2012, pp. 255ff.). Similar to woven fabrics, non-woven materials are finished to obtain the needed characteristics and geometries.

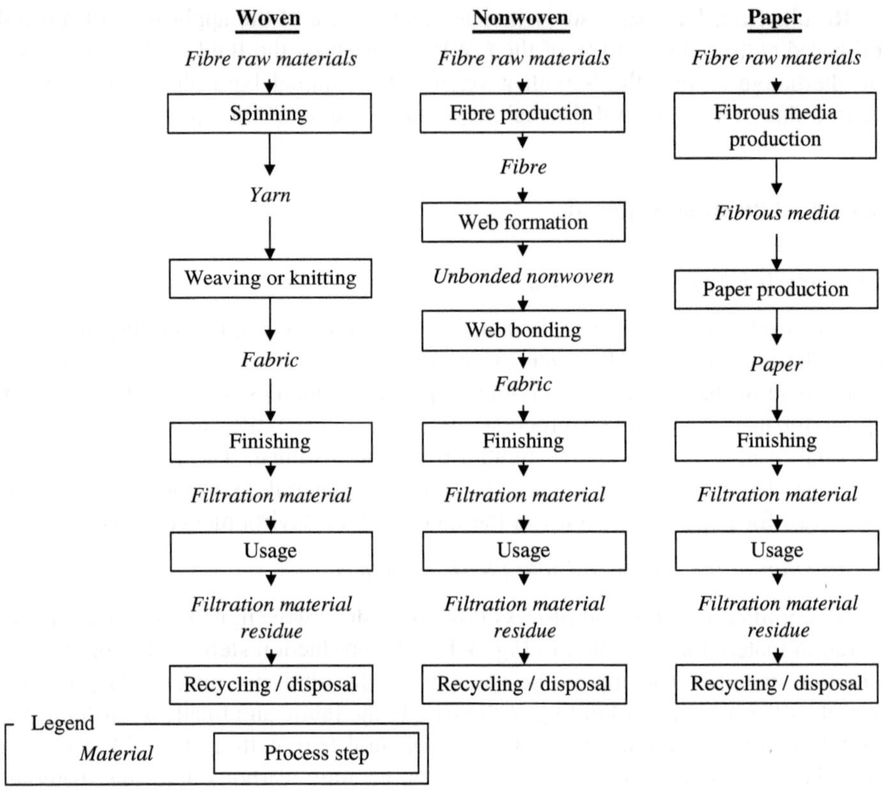

Fig. 4.42 Simplified material and process flow to produce paper, non-woven and woven filtration materials (According to Purchas and Sutherland 2002, pp. 35ff.; Fuchs and Albrecht 2012, pp. 33ff.)

The production of paper-based filtration material includes fibrous media production from natural (e.g. cellulose, wool) or artificial (e.g. thermoplastic polymers, glass) fibre raw materials to fibrous media. Afterwards, the fibres are mixed with water to form a suspension and filtered through a wire mesh to produce a thin layer of paper which can be compressed and dried (Purchas and Sutherland 2002, pp. 118f.). The finishing process for the paper used to produce filtration material also includes procedures to enhance or suppress certain physical and chemical properties, as well as generating the needed shape. Lastly, the filtration material is transported to the user. Depending on the type and design of the filtration material, different usage scenarios are possible. In case of a cutting fluid filter using an endless band, for example, the filtration material is reused several times, a similar process applies for cartridge filters or filter bags. When using non-endless band filters, the filtration material is used for a single time and then disposed of as hazardous waste. The filtration material is contaminated with cutting fluid, and as such it is classified as hazardous waste and incinerated.

LCA studies regarding the environmental assessment of textiles indicate that the production phase of non-woven structures from artificial fibre is connected with a high electrical energy-related environmental impact (Muthu 2014, pp. 13). A similar result can be assumed in the case of paper-based filtration materials (Dias et al. 2007, pp. 521ff.). As far as the disposal phase is concerned, the equivalent CO_2 impact is due to the incineration of the hazardous waste.

4.5 Evaluation

4.5.1 Single Technological, Economic or Environmental Evaluation

Figure 4.43 presents the visualisation of the single indicator impact (y-axis) depending on a single set of grinding process or system parameters (x-axis). The indicators used for representing the technological, economic and environmental indicators correspond with the indicators given in Fig. 4.6. Similar results apply for the grinding process and system parameters.

Further details about the application of single evaluation for the technological, economic and environmental evaluation are presented in the following paragraphs.

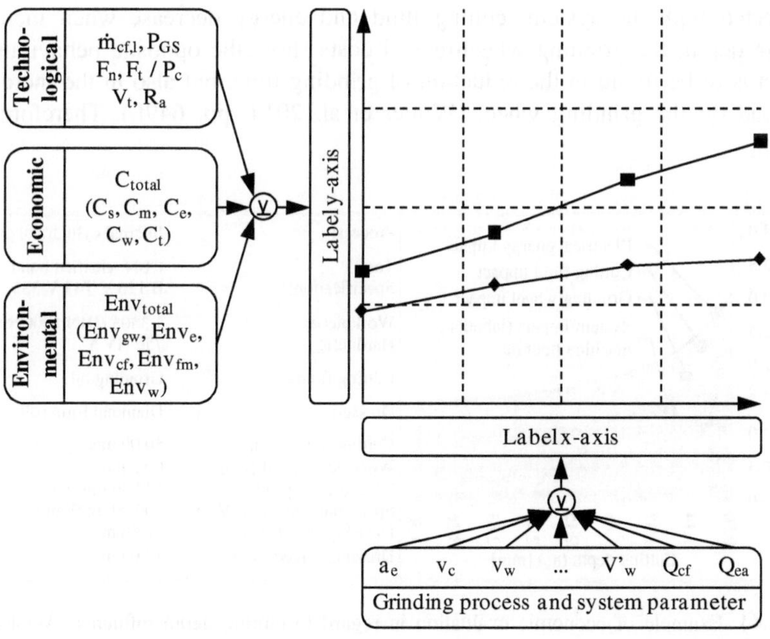

Fig. 4.43 Exemplary diagram: single technological, economic or environmental evaluation

Technological evaluation

Technological evaluation is the standard procedure used in production engineering to assess the relationship between technological indicators and process and system parameters the. The trend development of surface roughness, the forces and power demand connected with cutting depth and cutting speed is of major interest in the assessment of workpiece quality (see for example Fig. 4.19). This perspective can also be applied for the analysis of grinding wheel wear, cutting fluid loss or grinding system power demand patterns in relation to process and system parameters. The overall aim is usually that of maintaining the range of a specific indicator's value while reducing process time; this represents an indirect consideration of the economic perspective.

Economic evaluation

In case of the economic perspective, in particular process and system parameters are of interest, which results in minimal overall costs. As presented in Sect. 4.4.3, the overall costs include the sum of all expenses for the cutting fluid, grinding wheel, electrical energy, etc. However, each of those costs has a different behaviour depending on the process and system parameters being used. Figure 4.44 shows an example of economic evaluation of the influence of cutting depth on the costs related to electrical energy, cutting fluid, grinding wheel and grinding system (labour and machine-hour rate) (Winter et al. 2014, pp. 650).

The overall trend is dominated by system-related costs, followed by costs due to the use of grinding wheel, cutting fluid and electrical energy. While the costs connected with the system, cutting fluid and energy decrease when increasing cutting depth, the grinding wheel-related costs show the opposite behaviour. The reason is to be found in the reduction of grinding time, but also in the increase of the load on the grinding wheel (Winter et al. 2014, pp. 649f.). Therefore, it is

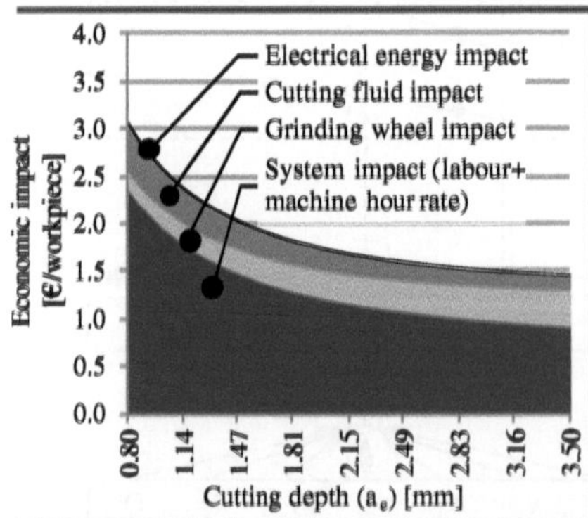

Process:	Internal cylindrical grinding
Tool: Specification:	CBN, vitrified bond B 126 V T43 V 385
Workpiece: Hardness:	1.3505 (DIN 100Cr6) 756 HV 5
Cutting fluid:	Grinding oil
Dresser:	Diamond form roll
Cutting speed (v_c):	50.00 m/s
Workpiece speed (v_w):	0.55 m/s
Trav. dress. speed (v_{fad}):	147.50 mm/min
Spec. mat. removed (V'_w):	100.00 mm³/mm
Dressing speed (v_d):	32.50 m/s
Dressing infeed (a_{ed}):	1.00 µm

Fig. 4.44 Example of economic evaluation in regard to cutting depth influence (Winter et al. 2014, pp. 650)

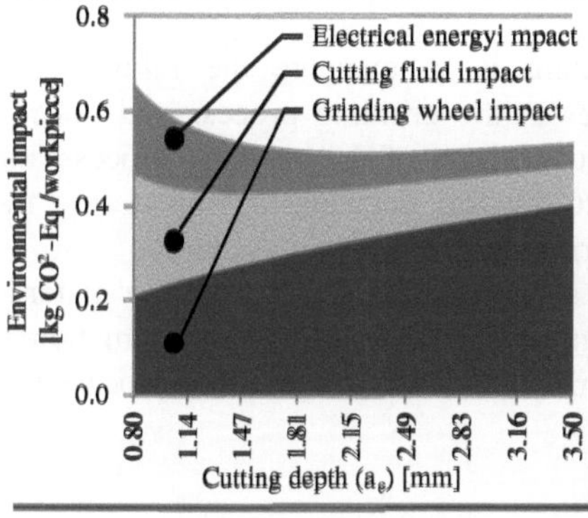

Process:	Internal cylindrical grinding
Tool: Specification:	CBN, vitrified bond B 126 V T43 V 385
Workpiece: Hardness:	1.3505 (DIN 100Cr6) 756 HV 5
Cutting fluid:	Grinding oil
Dresser:	Diamond form roll
Cutting speed (v_c):	50.00 m/s
Workpiece speed (v_w):	0.55 m/s
Trav. dress. speed (v_{fad}):	147.50 mm/min
Spec. mat. removed (V'_w):	100.00 mm³/mm
Dressing speed (v_d):	32.50 m/s
Dressing infeed (a_{ed}):	1.00 μm

Fig. 4.45 Example of environmental evaluation in connection with cutting depth influence (Winter et al. 2014, pp. 650)

helpful to present the overall costs not as a single trend, but rather as a stacked trend composed of different costs and their related impact.

Environmental Evaluation

Similar to the economic evaluation, the environmental evaluation focuses on the representation of the environmental impact in dependence with process and system parameters. Its aim is to identify the parameters which result in the lowest impact. The example in Fig. 4.45 presents the environmental impact due the demand of electrical energy, cutting fluid and grinding in dependence to the cutting depth. With increasing cutting depth, electrical energy and cutting fluid demand decreases as a result of declining process time, while the impact of the grinding wheel increases (Winter et al. 2014, pp. 650f.). The application of a stacked trend is a suitable option to evaluate environmental impact behaviour in conjunction with the parameters.

4.5.2 Integrated Evaluation

The single evaluation introduced in the previous section is a useful tool to analyse the influence of the process and system parameters. However, it does not take into account the enhancing and constraining interrelationships between the evaluation indicators. This limitation can be overcome by using the eco-efficiency approach according to DIN EN ISO 14045 and presented in Sect. 2.3.3.

As indicated in Fig. 2.17, the eco-efficiency assessment approach comprises five phases [goal and scope definition (1), environmental assessment (2), product system value assessment (3), quantification of eco-efficiency (4) and interpretation (5)]. These phases are described in the following sections on the basis of the requirements of the grinding process and system.

Goal and scope definition

The purpose of eco-efficiency assessment is to evaluate the influence of parameters as well as configuration changes in the grinding process and system. The intended audience are designers and supervisors of the grinding process and system. The evaluation is used in order to support decision-making during the definition as well as the operation of the grinding process and system.

The functional unit is the grinding of a specific volume of workpiece material featuring a specific geometrical and topographical accuracy. The system border includes the first, second and third hierarchical levels of the grinding process and system presented in Fig. 2.1.

Environmental assessment

A life cycle assessment according to ISO 14040 and 14044 is applied as an environmental assessment method. The impact indicator GWP (Env_{total} in units of kg CO_2-eq.) can be used to represent the environmental impact. Accordingly, the environmental assessment can be conducted by applying an LCA of the modelling indicators described in Fig. 4.6. The impact is obtained using the models described in Sect. 4.4.4 and related to the energy and resource flows described in Sect. 4.4.2.

Product system value assessment

The grinding process and system's values can either be represented by the technological or the economic impact categories. Surface roughness value in particular can be considered (e.g. R_a) a common technological impact indicator, due to its high industrial relevance among different surface integrity indicators and the comparable ease in its application. Alternatively, normal and tangential force as well as the demand of cutting fluid, grinding wheel or electrical power can be used instead. The impact indicators can be determined using measurements or prediction models (see Sect. 4.4.2). The overall costs/economic value of the grinding process and system (C_{total}) can be determined using the MFCA approach (see Sect. 4.4.3).

The decision on whether the technological or the economic value should be used depends on the selected grinding case. The technological value is appropriate if the focus on quality is prioritised. The opposite applies if monetary aspects are prioritised; in that case the economic value is used.

Quantification of eco-efficiency

The eco-efficiency indicator is defined, as shown in Eq. (2.9), by relating product system value and environmental system impact. The technological or economic value of the grinding process and system can be indicated using surface roughness (R_a) or the overall costs (C_{total}) (Eq. 4.72).

Both values are used in the following equations as enumerators, while the environmental impact (Env_{total}) is the denominator. Equation (4.92) formulates eco-efficiency with a technological focus (E/E_{tech}) and Eq. (4.93) formulates eco-efficiency with an economic focus (E/E_{econ}).

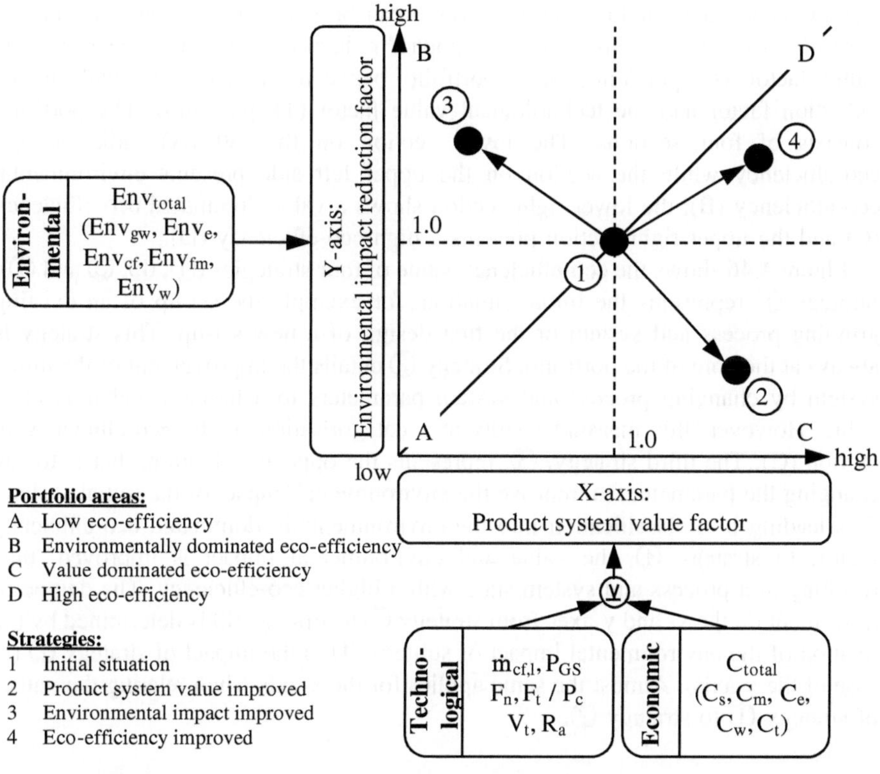

Fig. 4.46 Example of eco-efficiency portfolio for an integrated evaluation (according to DIN EN ISO 14045, 2012, pp. 35)

$$E/E_{\text{tech}} = \frac{\text{Technological value}}{\text{Environmental impact}} = \frac{R_{\text{a}}}{\text{Env}_{\text{total}}} \tag{4.92}$$

$$E/E_{\text{econ}} = \frac{\text{Economic value}}{\text{Environmental impact}} = \frac{C_{\text{total}}}{\text{Env}_{\text{total}}} \tag{4.93}$$

On these grounds, eco-efficiency is determined for the initial strategy as well as further strategies taking into consideration changed process and system parameters or configurations.

Interpretation

The interpretation step is based on eco-efficiency values quantified in the previous step. The values can be compared with each other with benchmarking using an eco-efficiency portfolio presented in Fig. 4.46 (according to DIN EN ISO 14045, 2012, pp. 35). The figure shows on the x-axis the product system value and on the y-axis the environmental impact factor. In case of product system value, either the technological value or the economic value can be taken into account, creating a portfolio between the environmental impact reduction factor and the economic value factor (EE-portfolio) or a portfolio between the environmental impact reduction factor and the technological value factor (TE-portfolio). The portfolio consists of four sections. The lower section on the left (A) indicates low eco-efficiency while the section on the upper left side presents environmental eco-efficiency (B); the lower right section shows a value-dominated eco-efficiency (C) and the upper right section presents a high eco-efficiency (D).

Figure 4.46 shows the eco-efficiency value of four strategies (①, ②, ③ and ④). Strategy ① represents the initial situation, for example the set-up of an existing grinding process and system or the first design of a new set-up. This strategy is always at the core of the portfolio. Strategy ② entails the improvement of the initial system by changing process and system parameters to achieve a higher product value. However, this measure results in a categorisation of the eco-efficiency in section (C). The third strategy, ③, represents the opposite situation, that is to say changing the parameters to improve the environmental impact of the initial system, thus leading to a classification into the environmentally dominated eco-efficiency sector. In strategy ④, the value and environmental impact is improved, thus resulting in a process and system state with a higher eco-efficiency. The degree of movement on the x- and y-axes from strategy ① to strategy ② is determined by the relation of the environmental impact of strategy ① to the impact of strategy ② (in case of the y-axis). Almost the same applies for the x-axis when relating the values of strategy ① to strategy ②.

4.6 Improvement

4.6.1 Improvement Planning

Improvement planning can be structured by focusing on the three main strategies for sustainable development (efficiency, consistency and sufficiency) (see Sect. 2.3.1). To this purpose, technological and organisational measures can be implemented in order to improve the grinding process and system. The exemplary measures presented in Fig. 4.47 are discussed in the following paragraphs and considered on the basis of their relationship with the three strategies of sustainable development.

The **efficiency** strategy focuses on the improvement of the output to input ratio and is a well-established option in manufacturing. Viable options in this context are technological measures comprising the variation of parameters and/or configurations and organisational measures including the change of contracts and maintenance procedures. Technological improvement measures comprise the following elements:

- The grinding process's and system's eco-efficiency can be influenced through the variation of process and system *parameters*. Either a single parameter or

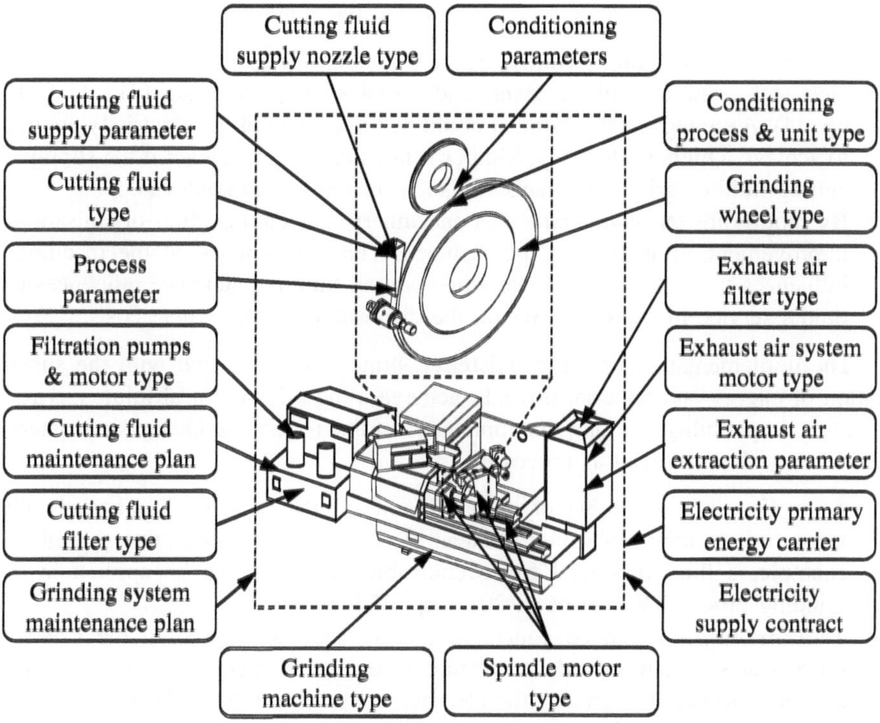

Fig. 4.47 Example of measures for the improvement of grinding processes and systems

alternatively multiple parameters can be modified. Relevant parameters are for example the grinding process parameters [cutting depth (a_e), cutting speed (v_c) and workpiece speed (v_w)], the conditioning parameters [dressing depth (a_{ed}), traverse dressing speed (v_{fad}) and dressing speed (v_d)] and system parameters [cutting fluid volume flow (Q_{cf}), exhaust air volume flow (Q_{ea})].

- Measures in the context of the variation of grinding system *configuration* can include the change of the grinding wheel and machinery. The machinery system, for instance, can be changed with regard to single components (e.g. spindle, drives) or whole sub-systems (e.g. grinding machine, filtration units).
- If either the *parameter* or the *configuration* is changed, it is important to make sure that the grinding system's configuration can fulfil the requirements of the selected process and system parameters and vice versa. If the cutting speed is increased, for example, then it must be checked whether the grinding wheel or the spindle drive can cope with a higher speed. As a result, it is worth taking into account strategies considering parameter and configuration modification to improve eco-efficiency.

Organisational improvement measures entail the change of:

- Within electricity supply *contracts,* the regulation of costs for electrical power, electrical work as well as fixed and variable standard costs (Thiede 2012, pp. 24). A strategy in this context could be the change of the electricity contract to achieve a higher efficiency. Similar actions can be conducted if the supply of cutting fluid or grinding wheels is also regulated within contracts.
- By modifying the *maintenance* of machinery and cutting fluid organisational improvement, strategies can be deduced. The maintenance of the machinery influences the machine-hour rate while cutting fluid maintenance influences the fluid's service life and, as a result, the disposal and replacement costs.

The implementation of the **consistency** strategy is connected with the substitution of energy and material flows to achieve a high degree of absolute environmental compatibility. Among the application strategies are the changes of material and energy flows as well as processes.

- An option in the context of *material flows* is the change of non-renewable substances to renewable ones. Instead of a mineral oil-based cutting fluid, for example, a fluid produced from renewable resources can be applied in the grinding process (Winter et al. 2013, pp. 254ff.).
- The *energy flows* can be considered, if the electricity used in the grinding process and system is produced from non-renewable energy carriers, and it can also be considered to change the electricity supply contract to be supplied with electricity produced from renewable energy carriers.

The application of the **sufficiency** strategy in industrial grinding processes and systems is constrained by a possible reduction of economic growth due to the deliberate abdication of business activities.

4.6.2 Improvement Procedure

The improvement procedure can be distinguished into a single-objective or multi-objective improvement approach. Within the single-objective procedure, either the economic, the environmental or the technological indicator is improved. However, due to conflicting relationships between these indicators (see Sect. 2.1.4), a single improvement leads to a unilateral solution. In order to achieve a Pareto-optimal grinding process and system, a multi-objective improvement procedure can be applied. This procedure identifies the minimal value of each indicator while considering the interdependency with the other indicators. At its foundation is the identification of a grinding process and system parameter combination which results into a Pareto-optimum (Winter et al. 2014, pp. 647).

For the achievement of this improvement procedure, a geometric programming algorithm is selected. This type of programming represents a class of nonlinear optimisations and can be used to solve convex optimisation problems. Furthermore, non-convex problems can also be transformed into convex problems (Boyd et al. 2007, pp. 67ff.). The improvement is performed by using the disciplined convex programming (cvx) package from the CVX Research Inc. for Matlab®. The form of the geometric program is presented in Eq. (4.94). The solving algorithm has the following structure:

$$\begin{array}{lll} \text{minimise} & f_K(x_l) & \\ \text{subject to} & g_i(x_l) \leq b_i, & i = 1, \ldots, p, \\ & g_i(x_l) = b_i, & i = p+1, \ldots, n, \\ & x_l \geq 0, & l = 1, \ldots, m \end{array} \tag{4.94}$$

where $f_K(x_l)$ is the target function as posynomial:

$$f_K(x_l) = \sum_{k=1}^{K} \left[w_k \cdot \left(K_l + \sum_{l=1}^{m} \left(C_l \cdot \prod_{i=1}^{n} x_i^{E_{li}} \right) \right) \right] \tag{4.95}$$

and g_i, $i = 1, \ldots, n$, are the constraint functions as monomials:

$$g_i(x_l) = K_l + \sum_{l=1}^{m} \left(C_l \cdot \prod_{i=1}^{n} x_i^{E_{li}} \right) \tag{4.96}$$

Also in the equation (Eq. 4.95) is w_k, which is the relative weight of the kth indicator, the additive constant (K_l), the multiplicative constant ($C_l > 0$), the exponential constants ($E_{li} \in R$), the values of the constraints (b_i, $i = 1, \ldots, n$) and the process parameter variables (x_l, $l = 1, \ldots, m$).

To reduce the calculation effort and complexity, the target function presented in Eq. (4.95) features a nonlinear formulation. Accordingly, empirical equations to measure the technological, economic and environmental indicators can be directly included within the target function. However, if the indicators are calculated using physical formulations, then the physical model has to be used to determine the indicator results for a set of multiple parameters. The results can then be used in a nonlinear formulation. This also applies for the constraint function. The target and constraint functions can also be formulated using the physical indicator equation, however, that would mean an increased calculation time.

The relative weight (w_k) can be obtained based on a weighted max–min model. The model measures the influence of the economic, environmental and technological indicators on eco-efficiency simultaneously. Therefore, the multiple indicators are normalised so that the ratio of the achievement levels is as close as possible to the ratio of the indicator weights (Lin 2004, pp. 409ff.). The weighted max–min model is formulated as follows:

$$\begin{aligned} \text{minimise} \quad & \sum_{k=1}^{K} w_k \cdot \frac{f_K(x_l) - \text{MIN}(f_K(x_l))}{\text{MAX}(f_K(x_l)) - \text{MIN}(f_K(x_l))} \\ \text{subject to} \quad & g_i(x_l) \leq b_i, \quad i = 1, \ldots, p, \\ & g_i(x_l) = b_i, \quad i = p+1, \ldots, n, \\ & x_l \geq 0, \qquad l = 1, \ldots, m \end{aligned} \tag{4.97}$$

where the relative weight (w_k) is of the kth indicator and $\sum_{k=1}^{K} w_k = 1$. The single-objective optimisations derive the maximal and minimal values of kth indicator as $\text{MAX}(f_K(x_l))$ and $\text{MIN}(f_K(x_l))$.

The relative weight can either be defined equally for each or a custom ratio between the technological, the economic and the environmental indicator can be defined based on the preferences of the decision-maker. A further option is an incremental varying of the weighting factor to perform a sensitivity analysis (Winter et al. 2014, pp. 652).

4.7 Application Cycle

4.7.1 Application Cycle and Software Environment

An application cycle to guide concept application is proposed in the following paragraphs, on the basis of the developed conceptual models introduced in the previous sections. When following the application cycle, users shall be supported in identifying purposefully, independently and systematically the eco-efficiency of the initial as well as the improved grinding process and system, while taking into consideration their specific requirements.

The application cycle is presented as a flow chart model in Fig. 4.48 and comprises nine steps. The steps are assigned to the proposed concept structure in Fig. 4.3 and include the following elements: description, modelling, evaluation and improvement. The application cycle follows the basic idea of the Plan, Do, Check, Act (PCDA) cycle also known as Deming cycle (Deming 2000, pp. 88). The user of the application is the designer and operator of the grinding process and system. The eventuality of an admin for the support of the user in the administration and maintenance of specific activities (modelling and validation) is also considered.

The application cycle is initiated with the aim of determining and improving the eco-efficiency of a new or existing grinding process and system. The first activity after the initiation is the goal and scope definition based on the requirements of the grinding process and system. This is followed by system description and configuration, which forms the basis of modelling in the next step. The results of the modelled system are then compared in a validation step with the results of a representative grinding process and system so as to maintain a high level of accuracy. Single evaluation visualises the single indicator impact (technological,

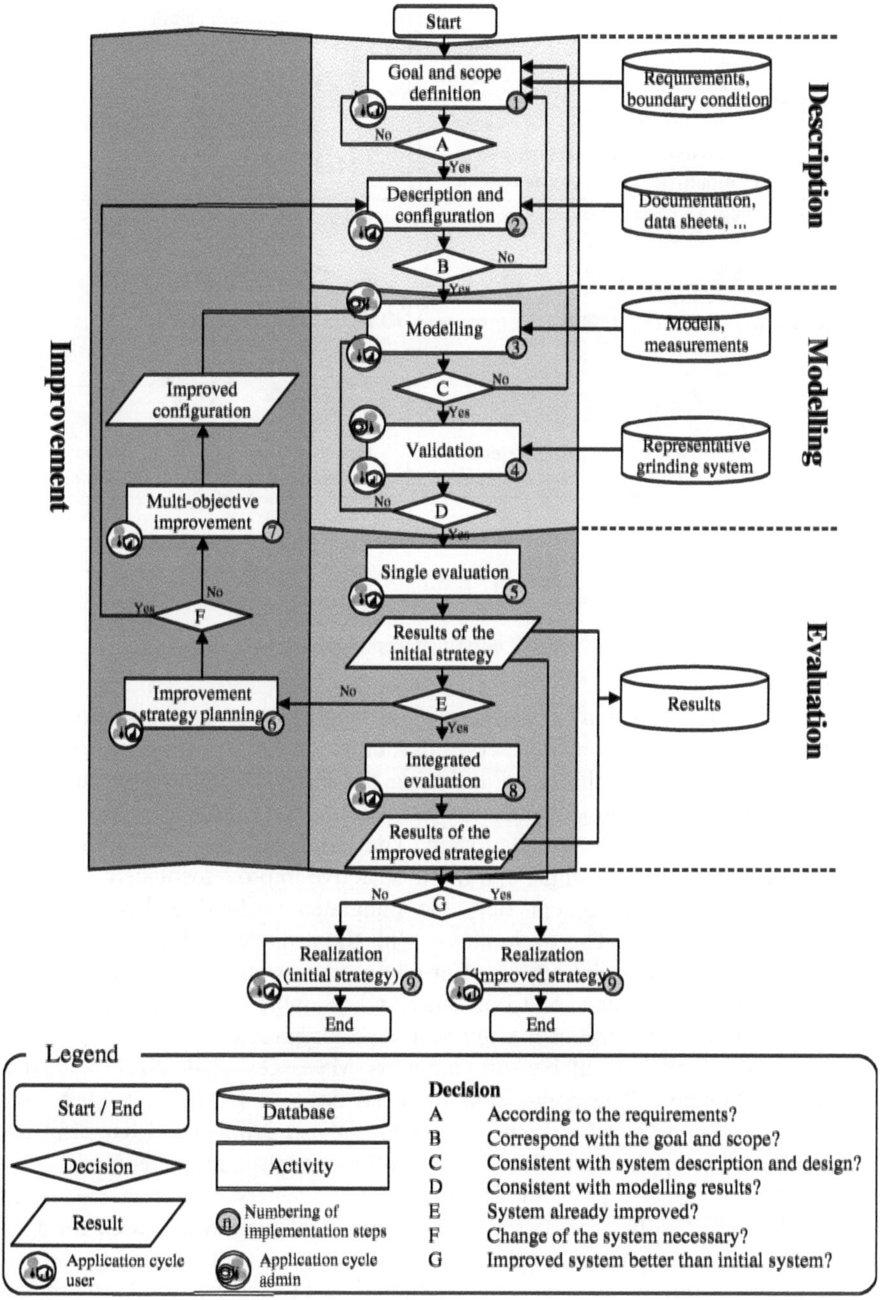

Fig. 4.48 Flow chart of the application cycle

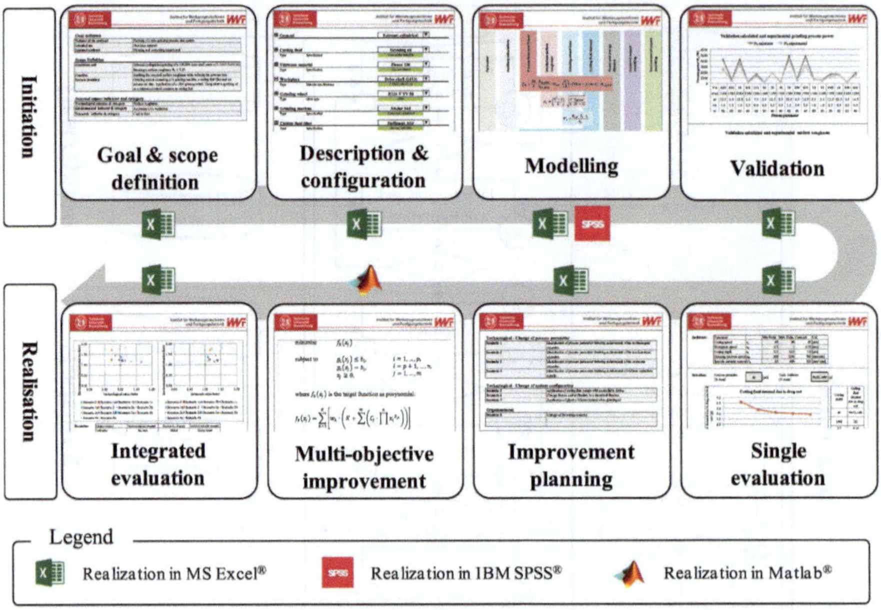

Fig. 4.49 Overview of software support during concept application

economic or environmental) depending on a single set of grinding process or system parameters. The evaluation serves as a basis for the following improvement planning, due to the possibility of analysing the influence of process parameter variation. Hence, improvement planning defines strategies, which are assumed to improve the grinding process and system. The initial and planned improvement strategies are analysed in a single and multi-objective improvement step. Based on the results of the improvement step, the integrated evaluation determines the strategy's eco-efficiency. The cycle closes with the achievement of either the initially planned strategy or the improved grinding process and system strategy.

The developed concept used to measure and to improve the eco-efficiency of grinding processes and systems is realised in a prototypical planning tool using the following commercially available software tools: MS Excel®, IBM SPSS® and The MathWorks Matlab®. The application of these three tools allows a purposeful realisation of the proposed concept combined with straightforward handling and practicability. The application of these three software tools is presented in Fig. 4.49 in connection with the application cycle.

The MS Excel® software is a spreadsheet-based calculation tool used for comprehensive mathematical and statistical calculations as well as the creation of graphical presentations for the results (Bildner 2011, pp. 9). Furthermore, the software tool has a high level of distribution. The software is used for data entry during the following steps: goal and scope definition, description and configuration, improvement planning and, during the calculation step within the modelling, validation and evaluation steps for the grinding process and system. For each of these application steps, a separate spreadsheet is provided to support the cycle. SPSS® by IBM is a comprehensive and powerful software tool used to analyse, model and forecast statistical data (Eckstein 2014, pp. 42). Due to those three characteristics (analysis, modelling and forecast), the software is used for nonlinear regression modelling in the modelling step. In addition to the provision of the regression coefficient, the software performs an analysis of variance to determine the predictive accuracy of the derived model. Multi-objective improvement is carried out using Matlab® by The MathWorks. The software can be used for complex and comprehensive numerical computations including linear algebra, interpolation, data accuracy, improvement (Venkateshan and Swaminathan 2014, pp. 9). Due to the ability to perform single- and multi-objective improvement computations, the software tool is used in the improvement step.

In the following sections, a detailed description of the nine steps is presented, including the input and outputs. In order to ensure a successful completion of the application cycle, the following assumptions are made:

- the user has general knowledge about the relationships occurring within the investigated grinding process and system,
- information with regard to the grinding process and system requirements is available,
- the user has access to measurement equipment (weight, surface roughness and electrical power) and software tools.

4.7.2 Steps of the Application Cycle

Step 1: Goal and scope definition

The goal and scope definition focuses on the requirements and the boundary conditions of the grinding process and system. Figure 4.50 summarises the needed input information and the provided output for the goal and scope definition step. The inputs are further described in the following sections.

The aim of goal definition was to describe as clearly as possible the reason for conducting the application cycle, the intended audience and the intended use of the results. The aim can take the form of increased eco-efficiency of an existing system or a higher eco-efficiency of a new grinding process and system, which needs to be

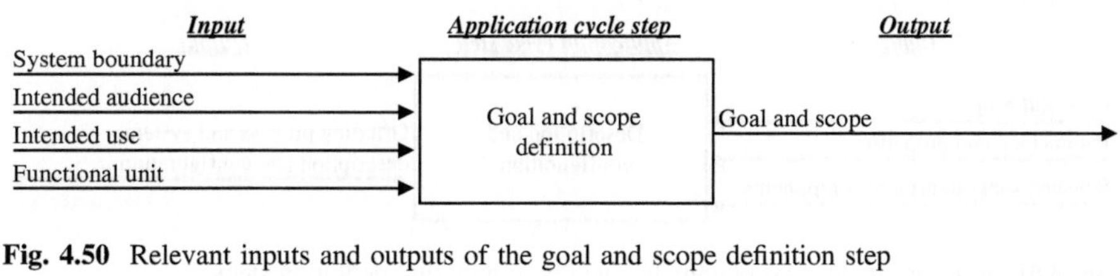

Fig. 4.50 Relevant inputs and outputs of the goal and scope definition step

configured and parameterised. An alternative to the integrated evaluation when measuring eco-efficiency is a single impact evaluation of an existing or new grinding process and system, which can be the purpose when realising the application cycle. The intended audience can be the designer and supervisor of the grinding process and system, who strive to reach a reconcilability between environmental impact and technological/economic value.

The functional unit and function of the grinding process and system need to be defined within the scope definition, together with system boundaries and limitations. The functional unit of the grinding process and system can be the production of a specific amount of workpieces while considering certain technological (e.g. geometrical and topographical accuracy), environmental (e.g. global warming potential impact) as well as economic (e.g. time, costs) functions and requirements. In this context, three different grinding strategies can be distinguished, focusing either on material removal rate, quality or the achievement of both (see Sect. 2.1). Accordingly, it is necessary to specify the selected environmental impact category if either a technological or economic value is used. The specification of the grinding process type is needed as well (e.g. surface grinding process, external or internal cylindrical grinding process). The specification of the selected grinding process type and grinding strategy should be as precise as possible, because both have a major influence on the grinding process' geometrical relationships, the selection of the grinding wheel and cutting fluid as well as the configuration of the machinery (e.g. size, required capability). Additionally, the system boundary has to be defined. It can include grinding machine, cutting fluid filtration and exhaust air filtration units as well as further equipment (e.g. workpiece changer, metring devices).

Before moving to the next step, it is necessary to check whether the requirements and boundary conditions are adequately formulated within the goal and scope definition. If this is not the case, the goals and scopes have to be modified.

Step 2: Description and configuration

The description and configuration, as described in Sect. 4.3, entails the characterisation of the grinding process and system level in connection with the process and system parameters and properties. The procedure used to describe and to configure the grinding process and system differs slightly depending on whether a new or an existing grinding process and system will be analysed. The starting point in both cases is the goal and scope definition. The needed input information and provided output information for the description and configuration steps are presented in Fig. 4.51. The input information is further described in the following paragraph.

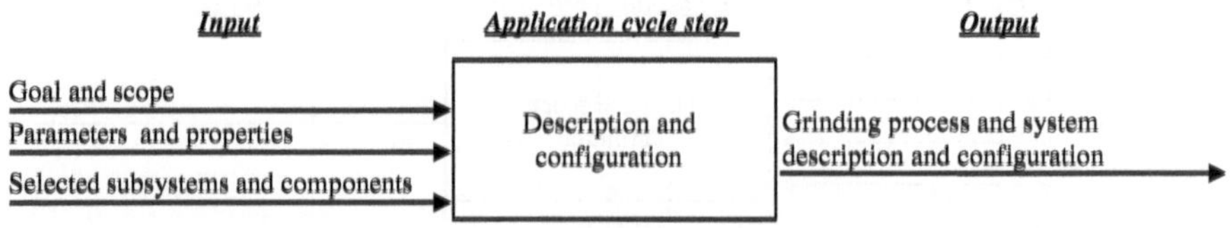

Fig. 4.51 Relevant inputs and outputs for the description and definition steps

The definition of the grinding process level entails the specification of workpiece, cutting fluid, grinding wheel and dresser, and the connected process and system parameters and properties (Fig. 4.4). In case of a new grinding process and system, a first selection of the process and system parameters has to be made based on experiences, recommendations of the manufacturers and literature. If an existing grinding process and system is to be analysed, then the given process and system parameters can be used. In order to gather the information required for the application cycle, a manual data entry form is presented in Appendix A.2.

The specification of the grinding system level includes, in accordance with Fig. 4.5, the machinery and boundary conditions. The machinery configuration includes the selected grinding machine, filtration units for cutting fluid and exhaust air as well as further equipment (e.g. workpiece changer, metring device). The boundary conditions include the specifications of the grinding system ambience (e.g. air temperature, air humidity), the prices for labour and electrical energy as well as the equivalent CO_2 impact of the electrical energy production. Connected with the machinery and boundary conditions are the corresponding properties. A manual data entry form for the grinding system level is presented in Appendix A.2.

The foundation of the description and configuration step is the availability of values for the specification of the properties. If a new grinding process and system is to be described/configured, the needed information can be gathered from product information, datasheets, manuals, literature, experience, recommendations of the manufacturer and operating standards. If an existing system needs to be analysed, then measurements can be conducted in addition to the aforementioned information. However, if it is still difficult to obtain the needed information—especially in the case of the description and configuration of a new grinding system—then possible options are given as follows, for example:

- **comparable data**: Electrical power demand measurements can be conducted with comparable machinery. Usually, a basic model (e.g. internal, universal, surface grinding machine) is selected from the product portfolio of the grinding machine's manufacturer. This basic model is then equipped with the needed components (e.g. spindle, conditioning unit). Almost the same applies for the decentralised filtration units. Assuming that the machinery manufacturer has conducted measurements or can conduct measurements, then the power demand in the idle state can be obtained from the machinery manufacturer. A similar approach can be conducted for cutting fluid and grinding wheel demand by using data of comparable grinding wheels and fluids.
- **given data**: On the basis of personal experience, experience of the provider of machinery, cutting fluid and grinding wheel and on historical data, a validation can be conducted. Another option in this context is the approximation of the electrical power demand in the idle state using the information given in electric circuit diagrams. Usually, the components used in the machinery, with their corresponding power demand, are presented in electric circuit diagrams. These values can be used to theoretically calculate the power demand. However, this procedure underlies the assumption that electrical components always operate

with a constant power demand and that all components are listed in the electric circuit diagrams. This procedure is difficult to implement for grinding wheel and cutting fluid demand.

Nonetheless, in order to obtain precise and reliable information, experiments still need to be conducted using the selected cutting fluid, grinding wheel and workpiece material.

To support the user during the description and configuration of the grinding system and process, Fig. 4.52 shows the execution of this application cycle step in an MS Excel® spreadsheet. The spreadsheet is structured into nine sections to allow a guided and separate input of the grinding process and system data. Either new data or existing data sets can be entered into the spreadsheet (the grinding wheel,

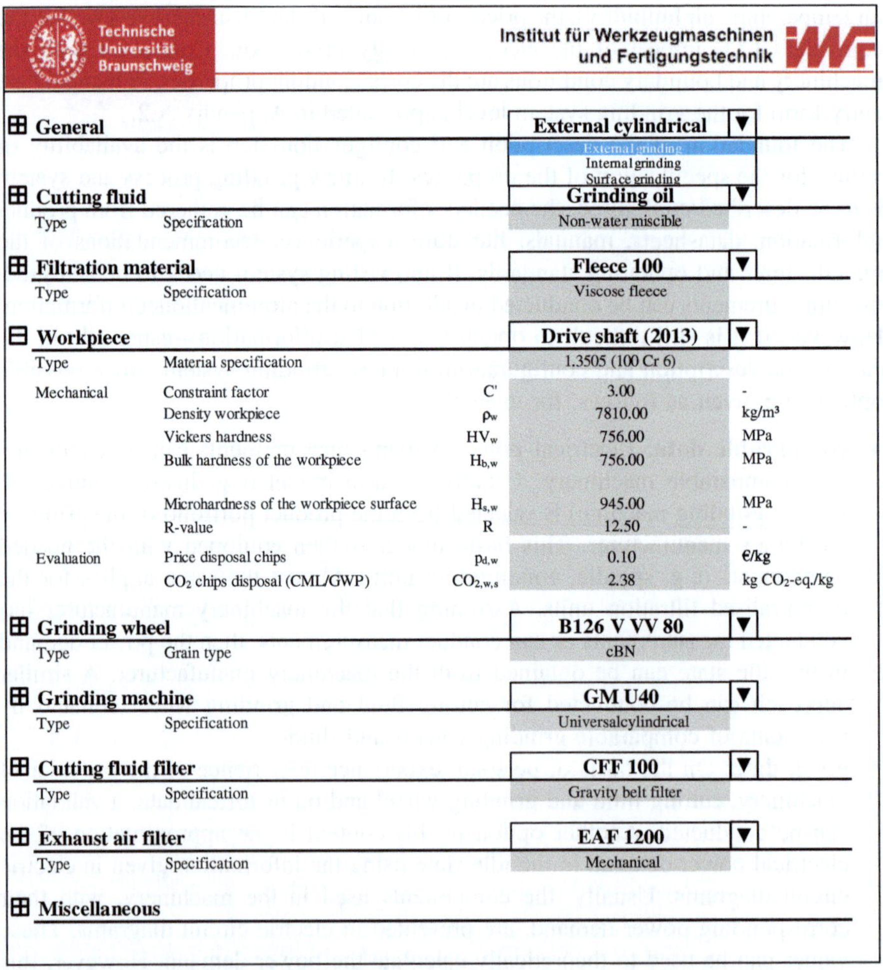

Technische Universität Braunschweig

Institut für Werkzeugmaschinen und Fertigungstechnik

⊞ **General**			**External cylindrical** ▼	
			External grinding / Internal grinding / Surface grinding	
⊞ **Cutting fluid**			**Grinding oil** ▼	
Type	Specification		Non-water miscible	
⊞ **Filtration material**			**Fleece 100** ▼	
Type	Specification		Viscose fleece	
⊟ **Workpiece**			**Drive shaft (2013)** ▼	
Type	Material specification		1.3505 (100 Cr 6)	
Mechanical	Constraint factor	C'	3.00	-
	Density workpiece	ρ_w	7810.00	kg/m³
	Vickers hardness	HV_w	756.00	MPa
	Bulk hardness of the workpiece	$H_{b,w}$	756.00	MPa
	Microhardness of the workpiece surface	$H_{s,w}$	945.00	MPa
	R-value	R	12.50	-
Evaluation	Price disposal chips	$p_{d,w}$	0.10	€/kg
	CO_2 chips disposal (CML/GWP)	$CO_{2,w,s}$	2.38	kg CO_2-eq./kg
⊞ **Grinding wheel**			**B126 V VV 80** ▼	
Type	Grain type		cBN	
⊞ **Grinding machine**			**GM U40** ▼	
Type	Specification		Universalcylindrical	
⊞ **Cutting fluid filter**			**CFF 100** ▼	
Type	Specification		Gravity belt filter	
⊞ **Exhaust air filter**			**EAF 1200** ▼	
Type	Specification		Mechanical	
⊞ **Miscellaneous**				

Fig. 4.52 Realisation of the step description and configuration in MS Excel®

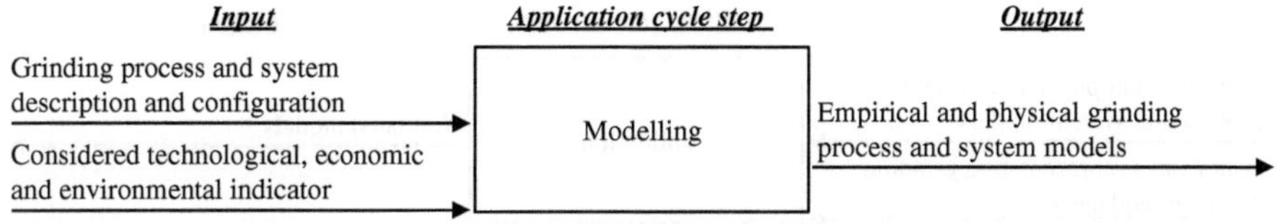

Fig. 4.53 Relevant inputs and outputs of the modelling step

cutting fluid, and so on can be selected, for example). These data sets represent the foundation of the following modelling and evaluation steps.

Before the next modelling step can be imitated, it is necessary to check whether the description and configuration of grinding process and system correspond with the goal and scope definition. If there is no match, the application cycle has to be restarted in order to refine either the goal and scope definition or the description and configuration. If there is a match, however, then the modelling step can be started.

Step 3: Modelling

The modelling step of the considered technological economic and environmental indicators (Fig. 4.53) can be conducted using the process and system parameters and properties specified in the previous step. The theoretical background and practical aspects of the modelling procedure were already introduced in Sect. 4.4. For the purpose of the modelling step, it is not important whether a new grinding process and system or an existing one is being modelled.

In order to minimise the modelling effort, only those models corresponding to the specifications included in the goal and scope definition need to be considered. This is important especially with regard to the selection of the technological or economic indicators used in the representation of the grinding process and system values. For example, if the economic value is represented on the basis of the economic indicator costs, then it is not necessary to measure/calculate the surface roughness value. In this case, the demand for power, cutting fluid and grinding wheel needs to be modelled together with the tangential force, as grinding power is calculated on the basis of the tangential force. If the economic value is represented using the technological value, then the surface roughness has to be modelled as well.

Due to the dependence of the modelling step on the specification made in the goal and scope definition, it is necessary to compare the models with the goal and scope before the validation step can be conducted. It is important to be consistent with the goal and scope definition; should this not be the case, then the cycle will have to be restarted.

Step 4: Validation

After the grinding process and system are properly defined, described and configured, as well as modelled, the validation can be conducted. Compared with

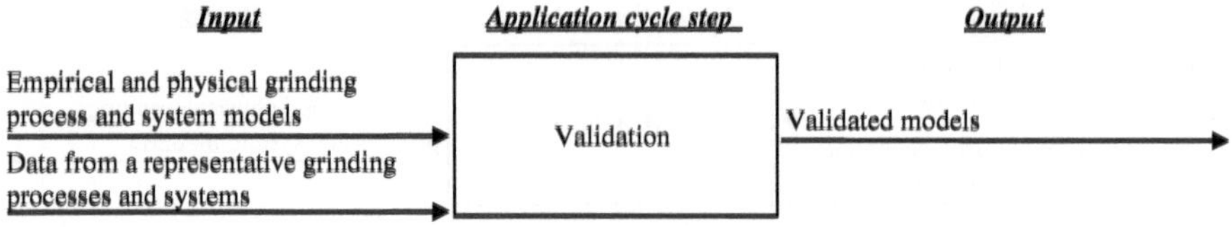

Fig. 4.54 Relevant inputs and outputs for the validation step

the numerous verifications conducted in Sect. 4.4, the validation ensures "that a model is an accurate representation of the real system" (Banks et al. 2004, pp. 310).

In order to conduct the validation step, data from a representative grinding process and system need to be used as input information in combination with the derived empirical and physical models (Fig. 4.54). This entails the measurement of the demand of power, cutting fluid and grinding wheel. For an existing grinding process and system, measurements can be conducted to ensure the validation. In case of a new grinding process and system, this procedure is not entirely applicable. However, for the purpose of conducting a validation measurement, data can also be approximated. Therefore, the options to obtain the needed information presented in step 2 can be applied.

The validation step represents an iterative process continuous comparison between the representative system data and the model. Arising deviations can be used to improve the model. If the model accuracy is acceptable, then the evaluation step can be conducted.

Step 5: Single evaluation

The single evaluation step comprises the derivation of diagrams and key figures, based on the means for single evaluation presented in Sect. 4.5.1. The needed input information and the provided output information for this step are presented in Fig. 4.55. Hence, indicators and parameters need to be selected for the presentation of the results of the grinding process and system. Single evaluation makes it possible to analyse the influence of process parameter changes and to identify the main contributing energy and material flows in regard to the environmental or economic impact. Hence, it serves as a foundation for improvement planning.

Lastly, it is necessary to check whether the grinding process and system have already been improved. If this is not the case, then improvement planning should be carried out.

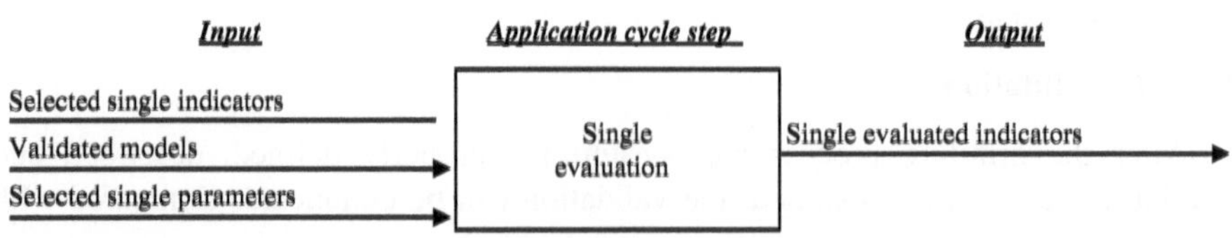

Fig. 4.55 Relevant inputs and outputs for the single evaluation step

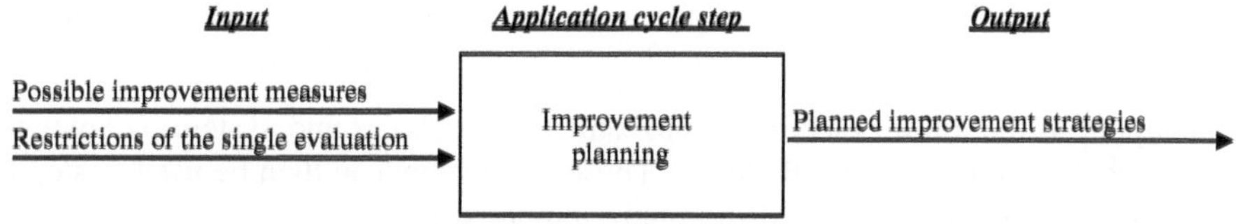

Fig. 4.56 Relevant inputs and outputs for the improvement planning step

Step 6: Improvement planning

This step is characterised by the planning of improvement strategies. Technological and organisational improvement measures are at the core of this step; they are described in Sect. 4.6.1 and allow an improvement of the eco-efficiency of the existing or new grinding processes and systems (Fig. 4.56). The improvement measures need to be selected by the designer or operator of the grinding processes and systems, considering the existing boundary conditions.

It is necessary to check, before the next step can be conducted, whether the described and configured grinding processes and systems need to be changed before considering the planned improvement strategies.

Step 7: Multi-objective improvement

This step is based on the application of the multi-objective improvement approach presented in Sect. 4.6.2. The improvement strategies obtained in the previous step, together with the initial system strategy, are both taken into consideration within the multi-objective improvement algorithm (Fig. 4.57).

The selection of how the value of the grinding process and system is represented to determine the eco-efficiency influences the multi-objective improvement as well. As far as the target function is concerned, in addition to the environmental indicator, either the technological or the economic value will have to be chosen. If a single evaluation instead of an integrated evaluation (with no eco-efficiency assessment) is conducted, then the economic, environmental and technological perspectives can be taken into consideration within the target function. In addition to the selection of the indicators, it is necessary to decide how the indicators are weighted. The options are as follows:

- equal weighting,
- a custom weighting favouring one indicator over the other(s) or

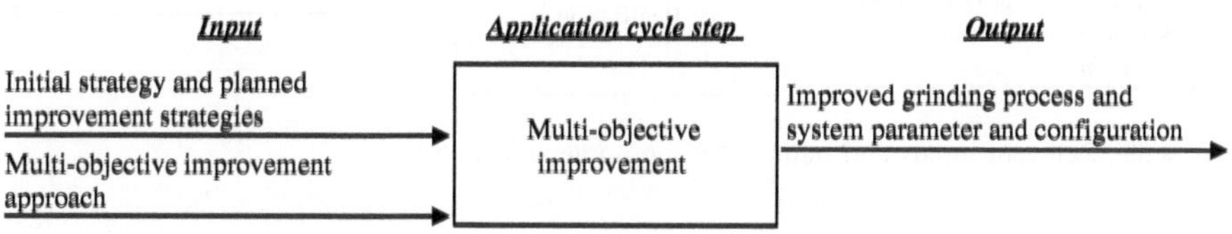

Fig. 4.57 Relevant inputs and outputs for the multi-objective improvement step

- an incremental weighting using a respective algorithm, in order to determine process and system parameter based on a sensitivity analysis.

Outputs of the multi-objective improvement are process and system parameters which result in a higher eco-efficiency. These parameters can then be used in step 8 to measure the eco-efficiency of the improvement strategy.

Step 8: Integrated evaluation

The integrated evaluation includes the eco-efficiency assessment method presented in Sects. 2.3.3 and 4.5.2. Due to the high degree of similarity between the eco-efficiency assessment method and this application cycle, the effort in conducting the five steps according to DIN EN ISO 14045 is reduced. Based on the selected system value, two different portfolios can be obtained. Either the technological value or the economic value in relation to the environmental impact reduction factor can be represented. Further inputs of the eco-efficiency portfolio are the results of the improvement strategies. The integrated evaluation serves as the foundation for the improvement strategy selection (Fig. 4.58).

Step 9: Realisation

Based on the results of the multi-objective improvement, it can be decided to realise (or maintain) either the initial system or the improved system (Fig. 4.59). This step represents a practical issue and is not focused on in detail. Regardless of the selected system, it is crucial to update the created models according to the decision to realise the process and system, to ensure an up-to-date standard for further eco-efficiency evaluations.

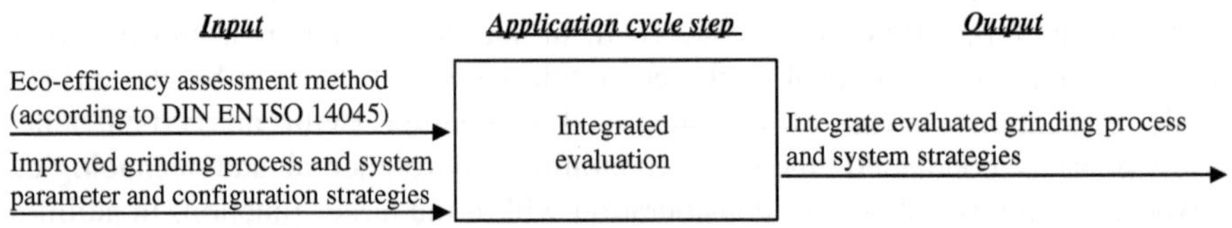

Fig. 4.58 Relevant inputs and outputs for the integrated evaluation step

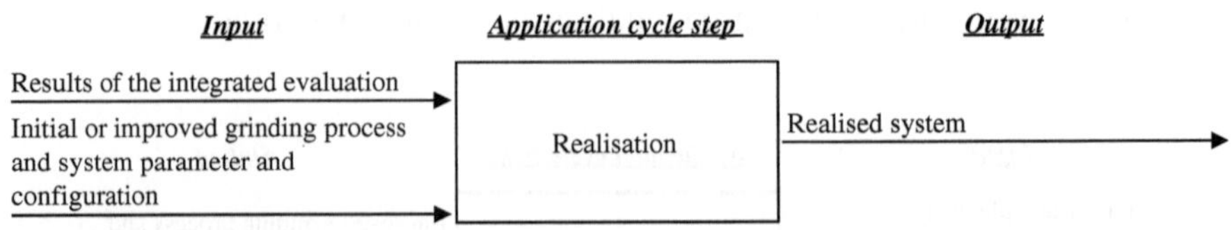

Fig. 4.59 Relevant inputs and outputs for the implementation step

References

Alduchov, O.A., Eskridge, R.E. (1996): Improved Magnus Form Approximation of Saturation Vapor Pressure. Journal of Applied Meteorology and Climatology, 35:601-609.

Ali, S.H. (2011): Ecological Comparison of Synthetic versus Mined Diamonds. Working paper, Institute for Environmental Diplomacy and Security, University of Vermont.

Archard, J.F. (1961): Single Contacts and Multiple Encounters. Journal of Applied Physics, 32/8:1420-1425.

Armstrong, J.S., Collopy, F. (1992): Error measures for generalizing about forecasting methods: Empirical comparisons. International Journal of Forecasting, 8/1:69-80.

Badger, J.A., Torrance, A.A. (2000): A comparison of two models to predict grinding forces from wheel surface topography. International Journal of Machine Tools and Manufacture, 40/8:1099-1120.

Banks, J., Carson II, J.S., Nelson, B.L., Nicol, D.M. (2004): Discrete-Event System Simulation. Fourth Edition. Prentice Hall, Upper Saddle River, USA.

Baumann, W., Herberg-Liedkte, B. (1996): Chemikalien in der Metallbearbeitung. Springer Verlag, Berlin, Germany.

Beverley, K.J. (2000): Evaporation of liquids from structured and non-structured mixtures. Ph.D Thesis, The University of Hull, Hull, UK.

Bierlich, R. (1976): Technologische Voraussetzungen zum Aufbau eines adaptiven Regelungssystems beim Außenrundschleifen. Dr.-Ing. Dissertation, RWTH Aachen, Aachen, Germany.

Bildner, C. (2011): Microsoft Excel 2010. Basiswissen. readersplanet, Passau, Germany.

Boyd, S., Kim, S.-J., Vanderberghe, L., Hassibi, A. (2007): A Tutorial on Geometric Programming. Optimization and Engineering, 8/1:67-127.

Büttner, A. (1968): Das Schleifen sprödharter Werkstoffe mit Diamant-Topfscheiben: unter besonderer Berücksichtigung des Tiefschleifens. Dr.-Ing. Dissertation, TU Hannover, Hannover, Germany.

Challen, J.M., Oxley, P.L.B. (1979): An explanation of the different regimes of friction and wear using asperity deformation models. Wear, 53/2:229-243.

Chen, X., Rowe, W.B. (1996): Analysis and simulation of the grinding process. Part II: Mechanics of grinding. International Journal of Machine Tools and Manufacture, 36/8:883-896.

Czenkusch, C. (2000): Technologische Untersuchungen und Prozessmodelle zum Rundschleifen. Dr.-Ing. Dissertation, Universität Hannover, Fortschrittsbericht, VDI Verlag, VDI-Reihe 2/530.

Davis, J.R. (1995): ASM Specialty Handbook Tool Materials. ASM International, Ohio, USA.

Deming, W.E. (2000): Out of the crisis. The MIT Press, Cambridge, USA.

Dettmer, T. (2006): Nichtwassermischbare Kühlschmierstoffe auf Basis nachwachsender Rohstoffe. Dr.-Ing. Dissertation, Technische Universität Braunschweig, Vulkan Verlag, Essen, Germany.

Dias, A.C., Arroja, L., Capela, I. (2007): Life Cycle Assessment of Printing and Writing Paper Produced in Portugal. The International Journal of Life Cycle Assessment, 12/7:521-528.

DIN 50320:1979-12 (1979): Verschleiß; Begriffe, Systemanalyse von Verschleißvorgängen, Gliederung des Verschleißgebietes. Beuth Verlag, Berlin, Germany.

DIN EN ISO 14045 (2012): Umweltmanagement - Ökoeffizienzbewertung von Produktsystemen - Prinzipien, Anforderungen und Leitlinien (ISO 14045:2012). Beuth Verlag, Berlin, Germany.

DIN EN ISO 14051 (2011): Umweltmanagement - Materialflusskosten-rechnung - Allgemeine Rahmenbedingungen (ISO 14051:2011). Beuth Verlag, Berlin, Germany.

DIN EN ISO 4287:2010-07 (2010): Geometrische Produktspezifikation (GPS) - Oberflächenbeschaffenheit: Tastschnittverfahren - Benennungen, Definitionen und Kenngrößen der Oberflächenbeschaffenheit. Beuth Verlag, Berlin, Germany.

Eckstein, P.P. (2014): Statistik für Wirtschaftswissenschaftler - Eine realdatenbasierte Einführung mit SPSS. Springer Gabler, Wiesbaden, Germany.

Fahrmeir, L., Künstler, R., Pigeot, I., Tutz, G. (2007): Statistik - Der Weg zur Datenanalyse. Springer Verlag, Berlin, Germany.

Fathima, K., Kumar, A.S., Rahman, M., Lim, H.S. (2003): A study on wear mechanism and wear reduction strategies in grinding wheels used for ELID grinding. Wear, 254/12:1247-1255.

Fingas, M.F. (2012): Studies on the Evaporation Regulation Mechanisms of Crude Oil and Petroleum Products. Advances in Chemical Engineering and Science, 2/2:246-256.

Fuchs, H., Albrecht, W. (Ed.) (2012): Vliesstoffe - Rohstoffe, Herstellung, Anwendung, Eigenschaften, Prüfung. WILEY-VCH Verlag, Weinheim, Germany.

GlaconChemie (2012): Your Green Chemistry Company. Produktinformation, Merseburg, Germany.

Grüntzig, A.W. (2011): Modeling of grinding wheel wear in electrolytic in-process dressing (ELID) grinding. Dr.-Ing. Dissertation, RWTH Aachen, Apprimus Verlag, Aachen, Germany.

Guinée, J. (Ed.) (2004): Handbook on Life Cycle Assessment - Operational Guide to the ISO Standards. Kluwer Academic Publishers, Dordrecht, Netherlands.

Guo, W. Jia, X. Guo, W.L, Xu, H.W., Shang, J., Ma, H.A. (2010): Effects of additive LiF on the synthesis of cBN in the system of Li3 N–hBN at HPHT. Diamond and Related Materials, 19/10:1296-1299.

Hansen, A., Hallmann, C., Schmehl, M. (2005): Produktökobilanz nichtwassermischbarer Kühlschmierstoffe auf Basis von Mineralöl, pflanzlichen Ölen sowie Altspeisefetten und technischen tierischen Fetten. DBU Abschlussbericht AZ 19122, Braunschweig, Germany.

Heuer, W. (1992): Außenrundschleifen mit kleinen keramisch gebundenen CBN-Schleifscheiben. Dr.-Ing. Dissertation, TU Hannover, Hannover, Germany.

Hokkirigawa, K., Kato, K., Li, Z.Z. (1988): The effect of hardness on the transition of the abrasive wear mechanism of steels. Wear, 123/2:241-251.

Hokkirigawa, K., Kato, T., Fukuda, T., Shinooka, M. (1998): Experimental and theoretical analysis of wear mechanism of metals in tilted block on plate type sliding. Wear, 214/2:192-201.

Ichida, Y., Fujimoto, M., Inoue, Y., Matsui, K. (2010): Development of High-Performance Vitrified Grinding Wheels using Ultrafine-Crystalline cBN Abrasive Grains. Journal of Advanced Mechanical Design Systems and Manufacturing, 4:1005-1014.

Kassen, G. (1969): Beschreibung der elementaren Kinematik des Schleifvorganges. Dr.-Ing. Dissertation, RWTH Aachen, Aachen, Germany.

Kent, J.A. (2007): Kent and Riegel's Handbook of Industrial Chemistry and Biotechnology. Springer Science + Business Media, New York, USA.

Klocke, K. (2009): Manufacturing processes 2 - Grinding, honing, lapping, Berlin, Germany.

Kloke, U. (2003): Auslegung von Bauteilreinigungsanlagen mit Hilfe eines Fachinformationssystems. Dr.-Ing. Dissertation, Universität Dortmund, Dortmund, Germany.

Klöpfer, W., Grahl, B. (2009): Ökobilanz (LCA) - Ein Leitfaden für Ausbildung und Beruf. WILEY-VCH Verlag, Weinheim, Germany.

Knop, M. (1989): Rechnergeführte Stellgrößenwahl beim Außenrundeinstechschleifen - Abrichtereinfluß und Temperaturberechnung. Dr.-Ing. Dissertation, RWTH Aachen, Aachen, Germany.

Kuhrke, B. (2011): Methode zur Energie- und Medienbedarfsbewertung spanender Werkzeugmaschinen. Dr.-Ing. Dissertation, Technische Universität Darmstadt, epubli, Berlin, Germany.

Kunsleben, A., Tschesche, J.R. (2010): Resource Cost Accounting (RKR) – A Synthesis of Business Management and Technology. Chemical Engineering & Technology, 33/4:589-592.

Li, W. (2012): Energy and eco-efficiency of manufacturing processes. Ph.D Thesis, The University of New South Wales, Sydney, Australia.

Li, W., Zein, A., Kara, S., Herrmann, C. (2011): An Investigation into Fixed Energy Consumption of Machine Tools. Proceedings of the 18th CIRP International Conference on Life Cycle Engineering, Technische Universität Braunschweig, Braunschweig, May 2-4, 268-273.

Lin, C.C. (2004): A weighted max-min model for fuzzy goal programming. Fuzzy Sets and Systems, 142/3:407-420.

Linke, B., Overcash, M. (2012): Life Cycle Analysis of Grinding. Proceedings of the 19th CIRP Conference on Life Cycle Engineering, University of California at Berkeley, Berkeley, USA, May 23-25, 293-298.

Linke, B.S. (2014): Sustainability concerns in the life cycle of bonded grinding tools. CIRP Journal of Manufacturing Science and Technology, 7:258-263.

Linke, K. (1992): Kennwerte keramisch gebundener Schleifscheiben. Dr.-Ing. Dissertation, TU Berlin, Carl Hanser Verlag, Berlin/ München, Germany.

LTA Lufttechnik GmbH (2014): Produkte, Nordrach, Germany.

Mackay, D., Matsugu, R.S. (1973): Evaporation rates of liquid hydrocarbon spills on land and water. The Canadian Journal of Chemical Engineering, 51:434-439.

Madanchi, N., Winter, M., Herrmann, C. (2015): Cutting fluid drag-out and exhaust air in grinding processes: Influence on the eco-efficiency. Proceedings of the 22th CIRP Conference on Life Cycle Engineering, The University of New South Wales, Sydney, Australia, April 7-9.

Malkin, S. (1968): The attritious and fracture wear of grinding wheels. Ph.D Thesis, Massachusetts Institute of Technology, Boston, USA.

Malkin, S., Guo, C. (2008): Grinding Technology: Theory and Applications of Machining with Abrasives. Industrial Press, New York, USA.

Mang, T., Dresel, W. (2007): Lubricants and Lubrication. WILEY-VCH Verlag, Weinheim, Germany.

Metzger, J.L. (1986): Superabrasive Grinding. Butterworth-Heinemann, Oxford, UK.

Montgomery, D.C. (2013): Design and analysis of experiments. John Wiley & Sons, Hoboken, USA.

Mortier, R.M., Fox, M.F., Orzulik, S.T. (2006): Chemistry and Technology of Lubricants. Springer Science + Business Media, Dordrecht, Netherlands.

Murray, V.R., Zhao, F., Sutherland, J.W. (2012): Life cycle analysis of grinding: a case study of non-cylindrical computer numerical control grinding via a unit-process life cycle inventory approach. Proceedings of the Institution of Mechanical Engineers, Part B: Journal of Engineering Manufacture, 226/10:1604-1611.

Muthu, S.S. (2014): Assessing the Environmental Impact of Textiles and the Clothing Supply Chain. Woodhead Publishing, Cambridge, UK.

Paulmann, R. (1994): Schleifen, Honen, Läppen: Grundlagen zu einem Verfahrensvergleich. Dr.-Ing. Dissertation, TU Braunschweig, VDI-Verlag, Braunschweig/ Düsseldorf, Germany.

Petuelli, G. (Ed.) (2002): Simulation des Kühlschmierstoffkreislaufs zur Optimierung einer umwelt- und ressourcenschonenden Produktionstechnik. DBU Abschlussbericht FZ 13581, Shaker Verlag, Aachen, Germany.

Plinke, W., Rese, M. (2006): Industrielle Kostenrechnung - Ein Einführung. Springer Verlag, Berlin, Germany.

Purchas, D.B., Sutherland, K. (2002): Handbook of Filter Media. Elsevier Advanced Technology, Oxford, UK.

Rabiey, M. (2010): Dry Grinding with cBN Wheels - The effect of structuring. Dr.-Ing. Dissertation, Universität Stuttgart, Jost-Jetter Verlag, Stuttgart, Germany.

Rowe, W.B. (2009): Principles of modern grinding technology. William Andrew, Elsevier, Oxford, UK.

Schenk, M., Wirth, S., Müller, E. (2014): Fabrikplanung und Fabrikbetrieb. Springer Verlag, Berlin, Germany.

Schleich, H. (1982): Schärfen von Bornitridschleifscheiben. Dr.-Ing. Dissertation, RWTH Aachen, Aachen, Germany.

Schlosser, R. (2013): Methodik zur Prognose der Nachhaltigkeit des Energie- und Stoffeinsatzes spanender Fertigungsprozesse. Dr.-Ing. Dissertation, RWTH Aachen, Shaker Verlag, Aachen, Germany.

Siniawski, M.T., Harris, S.J., Wang, Q. (2007): A universal wear law for abrasion. Wear, 262/7-8:883-888.

SKM Enviros (2011): Estimating the Energy Saving Potential from Small Motors and Machine Tools. Report on Machine Tools Research & Modelling, Manchester, UK.

Spiegel, P. (1994): Einsatzvorbereitung kunstharzgebundener CBN-Schleifscheiben. Dr.-Ing. Dissertation, Universität Kaiserslautern, Kaiserslautern, Germany.

Stiver, W., Mackay, D. (1984): Evaporation Rate of Spills of Hydrocarbons and Petroleum Mixtures. Environmental Science & Technology, 18/11:834-840.

Sutton, O.G. (1934): Wind Structure and Evaporation in a Turbulent Atmosphere. Proceedings of the Royal Society of London. Series A, Containing Papers of a Mathematical and Physical Character, 146/858:701-722.

Thiede, S. (2012): Energy Efficiency in Manufacturing Systems. Dr.-Ing. Dissertation, Technische Universität Braunschweig, Springer Verlag, Berlin, Germany.

Torrance, A.A., Badger, J.A. (2000): The relation between the traverse dressing of vitrified grinding wheels and their performance. International Journal of Machine Tools and Manufacture, 40/12:1787-1811.

Triemel J. (1976): Schleifen mit Bornitrid. Technischer Verlag Resch, Gräfelfing, Germany.

Umweltbundesamt (2000): Ökologische Bilanzierung von Altöl-Verwertungswegen. UBA-Texte 20/00, Berlin, Germany.

VDI 2089-1 (2010): Technische Gebäudeausrüstung von Schwimmbädern - Hallenbäder. VDI-Gesellschaft Bauen und Gebäudetechnik, Beuth Verlag, Berlin, Germany.

VDI 2884 (2005): Beschaffung, Betrieb und Instandhaltung von Produktionsmitteln unter Anwendung von Life Cycle Costing (LCC). VDI-Handbuch Betriebstechnik, Teil 4, Beuth Verlag, Berlin, Germany.

Vel, L., Demazeau, G., Etourneau, J. (1991): Cubic boron nitride: synthesis, physicochemical properties and applications. Materials Science and Engineering, 10/2:149-164.

Venkateshan, S.P., Swaminathan, P. (2014): Computational methods in engineering. Academic Press, Oxford, UK.

Verkerk, J. (1977): Presentation of the final report on the cooperative work on grinding wheel topography. CIRP Annals - Manufacturing Technology, 26/2:299-304.

Webster, J.A. (2007): Improving surface integrity and economics of grinding by optimum coolant application, with consideration of abrasive tool and process regime. Proceedings of the Institution of Mechanical Engineers, Part B: Journal of Engineering Manufacture, 221/12:1665-1675.

Wentorf, R.H. (1957): Cubic Form of Boron Nitride. The Journal of Chemical Physics, 26/4:956-956.

Winter, M., Bock, R., Herrmann, C. (2013): Investigation of a new polymer-water based cutting fluid to substitute mineral oil based fluids in grinding processes. CIRP Journal of Manufacturing Science and Technology, 6/4:254-262.

Winter, M., Ibbotson, S, Kara, S., Herrmann, C. (2015): Life cycle assessment of cubic boron nitride grinding wheels. Journal of Cleaner Production, 107:707-721.

Winter, M., Li, W., Kara, S., Herrmann, C. (2014): Determining optimal process parameters to increase the eco-efficiency of grinding processes. Journal of Cleaner Production, 66:644-645.

Winterthur Technology Group (2011): Precision grinding wheels - 2011 Catalogue.

Xie, Y., Williams, J.A. (1996): The prediction of friction and wear when a soft surface slides against a harder rough surface. Wear, 196/1-2:21-34.

Yegenoglu, K. (1986): Berechnung von Topographiekenngrößen zur Auslegung von CBN-Schleifprozessen. Dr.-Ing. Dissertation, RWTH Aachen, Aachen, Germany.

Zein, A. (2012): Transition Towards Energy Efficient Machine Tools. Dr.-Ing. Dissertation, Technische Universität Braunschweig, Springer Verlag, Berlin, Germany.

Zitt, U.R. (1999): Modellierung und Simulation von Hochleistungsschleifprozessen. Dr.-Ing. Dissertation, Universität Kaiserslautern, Kaiserslautern, Germany.

Chapter 5
Concept Application

Abstract This chapter focuses on the developed concept for the improvement of the eco-efficiency of grinding processes and systems, and its application within two case studies. To this purpose, the developed concept and application cycle presented in the previous chapter are applied in support of decision-making. The first case study focuses on the planning of a new external cylindrical grinding process (case study 1) of a fictitious but realistic company. The case study represents a scenario where different external cylindrical grinding processes and systems configurations can be compared on the basis of their eco-efficiency. The second case study focuses on the improvement of an existing (case study 2) internal cylindrical grinding process and system, as well as that of a fictitious but realistic company. The case study investigates a scenario which focuses on changes to the grinding process configuration. Both case studies represent a realistic decision scenario and are based on real data obtained from experiments and measurements.

5.1 Planning of a New Grinding Process and System

5.1.1 Initial Situation

A fictitious but realistic company producing bearings plans a new manufacturing plant in Germany. The company's aim is to plan and build a lean and green factory setting the benchmark for all manufacturing plants of the company with regard to economic and environmental indicators. Within this plant, different manufacturing processes shall be implemented, including the external grinding of hardened bearing rings made of DIN 1.3505 (100Cr6) featuring a hardness of 756 HV 5 (62 HRc). The designer of this external grinding process wants to compare different grinding system configurations, while achieving the lean and green aims defined by the management. To this purpose, the designer defines a basic strategy and conceives further improvement strategies at the grinding system level (different grinding machine types, decentralised or centralised cutting fluid and exhaust air

M. Winter, *Eco-efficiency of Grinding Processes and Systems*,
Sustainable Production, Life Cycle Engineering and Management,
DOI 10.1007/978-3-319-25205-6_5

filtration). Furthermore, changes in process parameters and cutting fluid are also considered. In addition to the requirement to achieve a lean and green grinding process and system, the designer also needs to take into account technological requirements connected with the achieved surface roughness and energy input into the workpiece surface. Therefore, the aim of the investigation is to measure the strategy's eco-efficiency and to implement the identified strategy, in order to achieve high eco-efficiency.

5.1.2 Concept Application

In the following, the application cycle from Sect. 4.7 is applied for the purpose of measuring the eco-efficiency of the grinding process and system to be planned.

Step 1: Goal and scope definition

The purpose of the assessment is to support decision-making with regard to the configuration of the external grinding process and system based on the eco-efficiencies of different configuration strategies. The intended audience are the planning, production and controlling departments, as well as all those actors interested and/or involved in the process. The results shall be used to support decision-making and present the strategy result to the intended audiences. The product system to be assessed is an external grinding process and system for manufacturing bearing rings. The scale of annual production ranges from 75,000 to 100,000 bearing rings, and the place of production is Germany.

The investigated function is the manufacturing of an external ground bearing ring made of DIN 1.3505 within a grinding system including a grinding machine, a cutting fluid filter and an exhaust air filter. The grinding process entails the application of a cBN grinding wheel and a cutting fluid. The functional unit is the production of one bearing ring featuring workpiece roughness $R_a \leq 0.35$ μm (the smaller, the better), specific normal force $F_n \leq 7.00$ N/mm and a specific grinding energy required for the material removal of a volume element of the bearing ring $e_c \leq 25.00$ J/mm^3.

The system boundary includes a gate-to-gate perspective within the manufacturing company (Fig. 5.1). The investigation considers the upstream chains for material extraction and production of the used energy and material flows (auxiliary goods) including grinding wheel, cutting fluid, filtration material and electrical energy as well as the end-of-life stage of these flows. The transport of the material and energy flows is excluded and so are the upstream and downstream chains of capital goods (machinery and equipment) and auxiliary resources (e.g. hydraulic fluid, grease). Furthermore, no allocation to adjacent systems is made.

The environmental impact of the considered auxiliary goods was calculated using the life cycle assessment method, taking into account the impact category global-warming potential into consideration, represented in equivalent carbon dioxide emissions. The product value is based on the functional value (represented

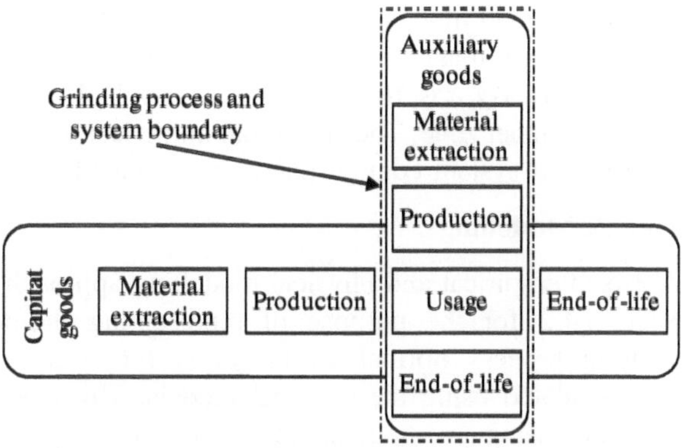

Fig. 5.1 Grinding process and system boundary

by the achieved workpiece roughness (R_a)) and the economic value (represented by the costs in Euro). Accordingly, an eco-efficiency indicator can be selected on the basis of either the technological value divided by the environmental impact or the economic value divided by the environmental impact. Both types of indicators were selected due to management requests for the purpose of a lean and green manufacturing plant/process.

Step 2: Description and configuration

The initial configuration of the grinding system includes a universal grinding machine, a decentralised cutting fluid filter and an exhaust air filter. The universal grinding machine (type GM U40) is able to perform external and internal cylindrical grinding operations. For the external cylindrical grinding operations, either a normal-speed or a high-speed spindle can be selected. The cutting fluid filter (type CFF 100) has a tank volume of 2000 l, uncontrolled supply drives and a gravity endless band filter is used. The decentralised exhaust air filter (type EAF 1200) is directly connected and controlled by the grinding machine and uses a mechanical filtration technique.

As far as the grinding process is concerned, the initial configuration entails the application of a cubic boron nitride grinding wheel and a mineral oil-based cutting fluid. The grinding wheel, with the B 126 V VV80 specifications, features a straight profile, a vitrified bond and a ceramic grinding wheel hub. The wheel has an external diameter of 400 mm and a grinding wheel width of 10 mm. The cutting fluid is grinding oil (mineral oil based) with a kinematic viscosity of 12 mm^2/s. The non-water miscible cutting fluid has an expected service life of 60 months. Due to the application of an endless band filter, it is assumed that the economic and environmental influence of the filtration material is negligible and thus will not be considered. The workpiece features a ring-like shape, which is made of DIN 1.3505 and has a hardness of 756 HV 5. During the grinding of the bearing ring, a specific material volume of 100 mm^3/mm is removed. The grinding swarf is disposed of via a hazardous waste incinerator. Almost the same applies for the cutting fluid. The selected process parameters for the initial description or strategy include a cutting depth (a_e) of 2.5 μm, a grinding wheel speed (v_c) of 40 m/s, a workpiece speed (v_w)

of 0.40 m/s and traverse dressing speed (v_{fad}) of 525 mm/min. The dressing speed (v_d) is directly related to grinding wheel speed by a ratio of 0.8. The cutting fluid volume flow (Q_{cf}) is 100 l/min and is supplied with a fluid pressure (p_{cf}) of 7 bar into the contact zone. The additional numerical values used to describe the grinding process and system configuration are presented in the Appendix A.3.

Step 3: Modelling

A mixed empirical and physical modelling approach is chosen in accordance with Sect. 4.4.2 for the purpose of representing the technological indicators. The grinding process' normal and tangential forces and the process power and energy are calculated using the physical models. The workpiece surface roughness was measured using an empirical model. This also applies for the calculation of grinding wheel wear. The economic and environmental impact indicators are determined according to the modelling approaches presented in Sects. 4.4.3 and 4.4.4.

Step 4: Validation

A second-order composite experimental design was used to validate the modelled values. To this purpose, the influence of the cutting speed, workpiece speed, cutting depth, traverse dressing speed and specific volume of removed material were all taken into account. The details of the experimental design are presented in Table 5.1. Based on this design for experiments, grinding trails were conducted on a comparable grinding system, but using the planned grinding wheel and cutting fluid.

Overall, 52 experiments with varying process parameters were conducted. During the grinding trials, the achieved workpiece roughness, the tangential force, the normal force, the grinding power and the tool wear volume were measured. The validation results for grinding process power and achieved workpiece roughness are presented in Fig. 5.2. The upper part of the figure shows the experimental conditions, the centre part presents the validation results for grinding power, and the lower part shows the results for workpiece roughness.

The first analysis indicates a good match between the measured and calculated values. When comparing the influence of the process parameters on grinding power, the cutting depth stands out as having a great influence, while the traverse dressing speed has an impact on workpiece roughness. These results match the relationships presented in Sects. 4.4.2.2 and 4.4.2.3. In the case of the grinding force, especially in the first 17 experimental trials, a higher deviation between the observed and calculated data is evident. This can be explained as a result of the very finely

Table 5.1 Parameters for the external grinding experimental design

Process parameters	Minimum	Mean	Maximum
Cutting depth, a_e [µm]	2.5	7.5	12.5
Workpiece speed, v_w [m/s]	0.5	0.75	1.00
Cutting speed, v_c [m/s]	40.00	60.00	80.00
Spec. volume removed, V'_w [mm³/min]	50.00	325.00	600.00
Traverse dressing speed, v_{fad} [mm/min]	138.90	764.00	1389.00

Cutting speed (v_c [m/s]):	Varying	Process:	External cylindrical grinding
Workpiece speed (v_w [m/s]):	Varying	Tool:	CBN, vitrified bond
Cutting depth (a_e [µm]):	Varying	Specification:	B126 V VV 80
Trav. dress. speed (v_{fad} [mm/min]):	Varying	Workpiece:	1.3505 (DIN 100Cr6)
Spec. mat. removed (V'_w [mm³/mm]):	Varying	Hardness:	756 HV 5
Dressing speed (v_d [m/s]):	Varying	Cutting fluid:	Grinding oil
Dressing infeed (a_{ed} [µm]):	1	Dresser:	Diamond form roll

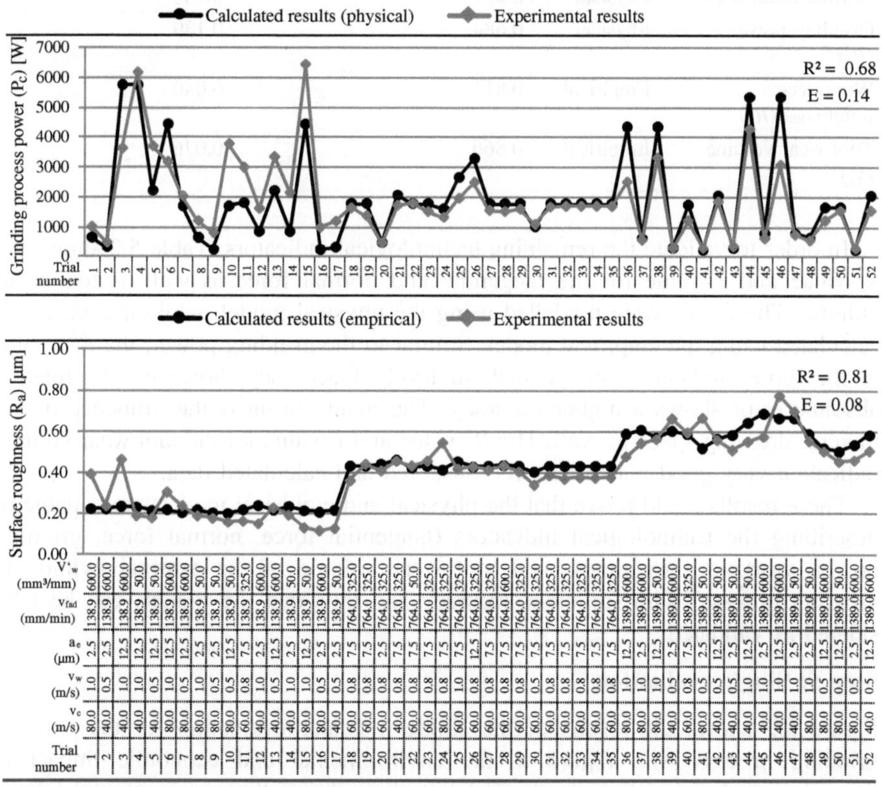

Fig. 5.2 Validation of the technological indicators

dressed grinding wheel surface, due to a low traverse dressing speed. The deviation increases with rising cutting depth. However, with increasing traverse dressing speed, the deviation between measured and observed data increases. Modifications of the physical formulation of the grinding power only lead to minor improvements. The application of the coefficient of determination (R^2) and of the relative absolute error (E) shows an ambiguous result in case of the grinding power. The R^2-value is comparably low, which indicates a medium level of accuracy, yet the E-value indicates a comparably higher accuracy, due to a low value. The reason for the low R^2-value is the deviation between the first 17 data points; without these data, the R^2-value would increase to 0.924. Therefore, it can be assumed that the physical model fits well. A similar result applies for the empirical model, in which R^2-value and E-value indicate a comparably high level of accuracy.

Table 5.2 Evaluation of predictive accuracy of the technological indicators

Technological indicator	Model type	Coefficient of determination (R^2) [−]	Relative absolute error (E) [−]
Tangential force (F_t)	Physical	0.607	0.174
Normal force (F_n)	Physical	0.631	0.182
Grinding power (P_c)	Physical	0.684	0.139
Workpiece roughness (R_a)	Empirical	0.811	0.080
Tool wear volume (V_t)	Empirical	0.866	0.076

In order to validate the remaining technological indicators, Table 5.2 shows the R^2-value and E-value for the tangential and normal force as well as tool wear volume. The forces were modelled using the physical model, while the wear was calculated using the empirical model. Similar to the grinding power, the R^2-values of the forces indicates only a medium level of accuracy; however, the relative absolute error shows a higher accuracy. The main reason is the influence of the traverse dressing speed as well. The R^2-value and E-value for the tool wear volume indicate a very good match between observed and calculated data.

These results could prove that the physical and empirical models are capable of describing the technological indicators (tangential force, normal force, grinding power, workpiece roughness and tool wear volume of the external cylindrical grinding process) with sufficient accuracy and thus that the models can be considered as validated.

Step 5: Single evaluation

Figure 5.3 shows the results of the single indicator evaluation for the external grinding process with regard to the environmental and economic impact in dependence of the cutting depth. The left chart shows the environmental impact connected with the demand of grinding wheel, cutting fluid, electrical energy as well as material disposal. The right chart presents the system, energy, material and disposal costs. In addition to the varying cutting depth, the remaining process parameters are kept constant. The cutting depth was selected due to its major influence on the environmental and economic impact. Workpiece speed could be selected as well instead of cutting depth. Cutting speed and traverse dressing speed have little influence on the economic and environmental indicators.

With increasing cutting depth, the economic and the environmental impact decrease. The reason is the reduction of cutting time and the time-dependent demand of cutting fluid, electrical energy and grinding system. As far as environmental impact is concerned, electrical energy demand in particular has a high influence, followed by cutting fluid, grinding wheel demand and material disposal. In the case of economic impact, the expenses due to the system (including labour and machine-hour rate) have the highest impact, followed by material, energy and disposal costs.

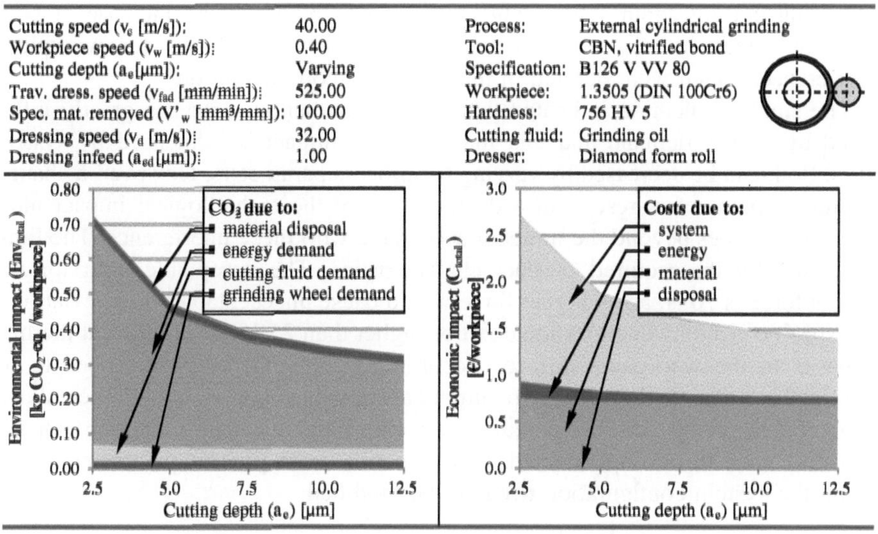

Fig. 5.3 Economic and environmental evaluation of the external grinding process

Based on the results presented in Fig. 5.3, it can be assumed that a high cutting depth should be selected. However, when considering the technological impact indicators, specific grinding energy and normal force rather than a high cutting depth have a negative influence (Fig. 5.4). The figure shows on the right side the specific grinding energy and on the left side the normal force measured against an

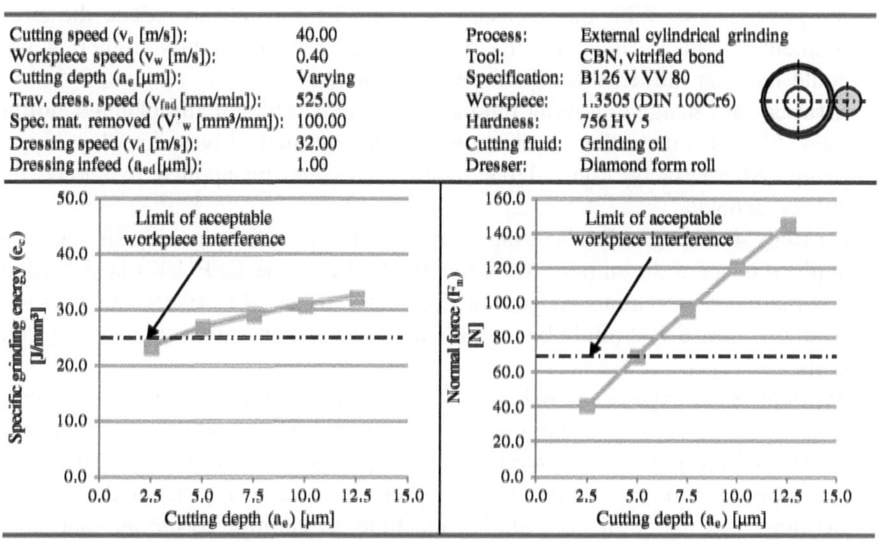

Fig. 5.4 Technological evaluation of the external grinding process

increasing cutting depth. The rising cutting depth results in an increase of both indicators. This increase extends beyond the acceptable limits for workpiece interference, which were defined in the goal and scope definition.

The single evaluation indicated that the environmental impact is greatly influenced by energy demand and that the economic impact is influenced by system costs. Both can be decreased by varying the process parameters. However, a change in process parameters results in a deterioration of the technological impact indicator, which goes beyond the limits of acceptable workpiece interference. The limit is defined by the required product characteristics. The acceptable limit for the normal force is 70 N, and a higher force would cause unacceptable form deviations. The specific grinding energy should not be higher than 25 J/mm^3 to prevent thermal damages to the workpiece surface. Accordingly, process parameters need to be identified in a manner to meet the requirements of the economic, environmental and technological perspectives. To reduce the environmental and economic impact, measures need to be implemented, independently from process parameters. As a result, the system configuration needs to be modified.

Step 6: Improvement planning

The single indicator evaluation showed that the grinding process and system could be improved through the modification of process parameters and grinding system configuration. Accordingly, the proposed improvement strategies are as follows:

- **Strategy 1**: Change of process parameters. The change of the process parameters can be further divided according to the selection of parameters, which either foster technological impact (1a), environmental impact (1b), economic impact (1c) or equally weighted impact (1d).
- **Strategy 2**: Change of the cutting fluid. A glycerol dilution is applied instead of the grinding oil. With this measure, it might be possible to reduce the environmental impact. However, the procurement price is slightly higher, which could have an influence on the economic impact.
- **Strategy 3**: Change of cutting fluid supply pumps. The initially planned cutting fluid filter features the application of uncontrolled drives for the supply pump. Alternatively, controllable pump drives can be applied to reduce the energy demand.
- **Strategy 4**: Change of the cutting fluid filter. In the initial planning strategy, the application of a decentralised cutting fluid filter (type CFF 100) is proposed. Alternatively, a centralised cutting fluid filter (type CFF 1000) can be used. With this measure, it might be possible to reduce the energy demand and system costs.
- **Strategy 5**: Change of the exhaust air filter. Similar to the cutting fluid filter, a centralised exhaust air filtration system (type EAF 12,000) can be applied instead of the current decentralised system (type EAF 1200). An advantage could be the reduction of the machine-hour rate and energy demand.
- **Strategy 6**: Change of the grinding machine. In addition to the auxiliary systems, it would be worth investigating which advantages would arise if a

specialised grinding machine (type GM E22) was to be applied instead of the currently planned universal grinding machine (type GM U40).

- **Strategy 7**: Combined change. Based on the result of the previous strategies, a combined change can be conducted leading to an overall improved grinding process and system. Within the strategy (7a and c), the grinding oil is applied, while in strategy (7b and d) the glycerol dilution is used instead.

Only those parameters or configurations addressed in the strategy have been changed. The remaining parameters or configuration details are kept unchanged. The sequence of the proposed improvement strategies arises from the effort of conducting the change. In the case of the process parameters, only slight modifications are necessary; this also applies in the case of the cutting fluid and supply pump drives. However, the effort increases, while for system components, the effort is significantly higher.

Step 7: Multi-objective improvement

The results of the single- and multi-objective improvement are presented in Table 5.3 for grinding oil application and in Table 5.4 for glycerol dilution application. The tables show the improved process parameters with regard to the respective improvement focus. The improvement focus includes environmental impact, economic impact, technological impact and an equally weighted economic, technological and environmental impact. The process parameters include the

Table 5.3 Parameters resulting in a grinding process improvement (application of grinding oil)

Grinding process parameters	Improvement focus				
	Initial	Environmental	Economic	Technological	Equal
Grinding speed, v_c [m/s]	40.00	80.00	80.00	80.00	80.00
Workpiece speed, v_w [m/s]	0.40	0.60	0.60	0.40	0.53
Cutting depth, a_e [μm]	2.5000	6.4077	6.4077	2.5000	6.6641
Traverse dressing speed, v_{fad} [mm/min]	525.00	551.02	551.02	100.00	100.00

Table 5.4 Parameters resulting in a grinding process improvement (application of glycerol dilution)

Grinding process parameters	Improvement focus				
	Initial	Environmental	Economic	Technological	Equal
Grinding speed, v_c [m/s]	40.00	80.00	80.00	80.00	80.00
Workpiece speed, v_w [m/s]	0.40	0.40	0.40	0.40	0.40
Cutting depth, a_e [μm]	2.5000	9.0251	9.0251	2.5000	7.5119
Traverse dressing speed, v_{fad} [mm/min]	525.00	470.78	470.78	100.00	100.00

cutting speed, the workpiece speed, the cutting depth and the traverse dressing speed. The specific volume of material removed is assumed constant ($V'_w = 100\ \text{mm}^3/\text{mm}$). During the single- and multi-objective improvement, the constraints are with regard to workpiece surface roughness, normal force and specific grinding energy apply, as described in the goal and scope definition.

In both cases, the single indicator improvement for the environmental and economic impact results in the same process parameters. This is due to the shared data basis for the demand of time, cutting fluid, electrical energy and tool. The parameters reducing these demands consequently reduce the connected environmental and economic impact. The cutting depth, workpiece speed and traverse dressing speed are considerably high, resulting in a short process and conditioning time. The parameters would normally even be higher; however, the constraints limited this potential increase. The single-objective improvement for the technological impact results in different process parameters. When these parameters are applied, then a surface roughness of 0.19 μm could be achieved; however, the economic impact would be 67 % and the environmental impact 111 %, higher than the respective improvement strategy. The reason is the result of the very low cutting depth and workpiece speed, resulting in a long process time as well as material and energy demand. The equal focus on the economic, environmental and technological perspective in the multi-objective improvement results in a Pareto-optimal process parameter configuration.

The comparison between the application of the grinding oil and the glycerol dilution results in different process parameters (Tables 5.3 and 5.4). The reasons are the different physical characteristics and capabilities of both fluids. When comparing both fluids and considering the process parameters from a multi-objective improvement focus, the application of the grinding oil leads to a lower process time by 7.5 % and to reduced overall costs by 4.2 %. However, the overall environmental impact is about 4.4 % higher, compared with the application of grinding oil.

Step 8: Integrated evaluation

Figure 5.5 shows the integrated evaluation for the external cylindrical grinding process. The eco-efficiency portfolio includes the results for the 14 considered strategies. The upper graph shows the portfolio between the environmental impact reduction factor and the economic value factor (EE-portfolio), while the lower graph shows the portfolio of the environmental impact reduction factor and the technological value factor (TE-portfolio).

In the first strategy, the process parameters are modified. The economically (1a) and environmentally (1b) oriented improvement results in a significantly higher eco-efficiency in the EE-portfolio. As far as the TE-portfolio is concerned, the strategy (1a and b) also result in an environmental improvement. However, the technological value increases slightly. The opposite situation applies to strategy (1c). The technological improvement focus results in a high technological value, a marginally higher economic value and a smaller environmental impact reduction factor compared to the initial situation. The equally weighted multi-objective strategy (1d) leads to a higher eco-efficiency in the EE- and the TE-portfolios.

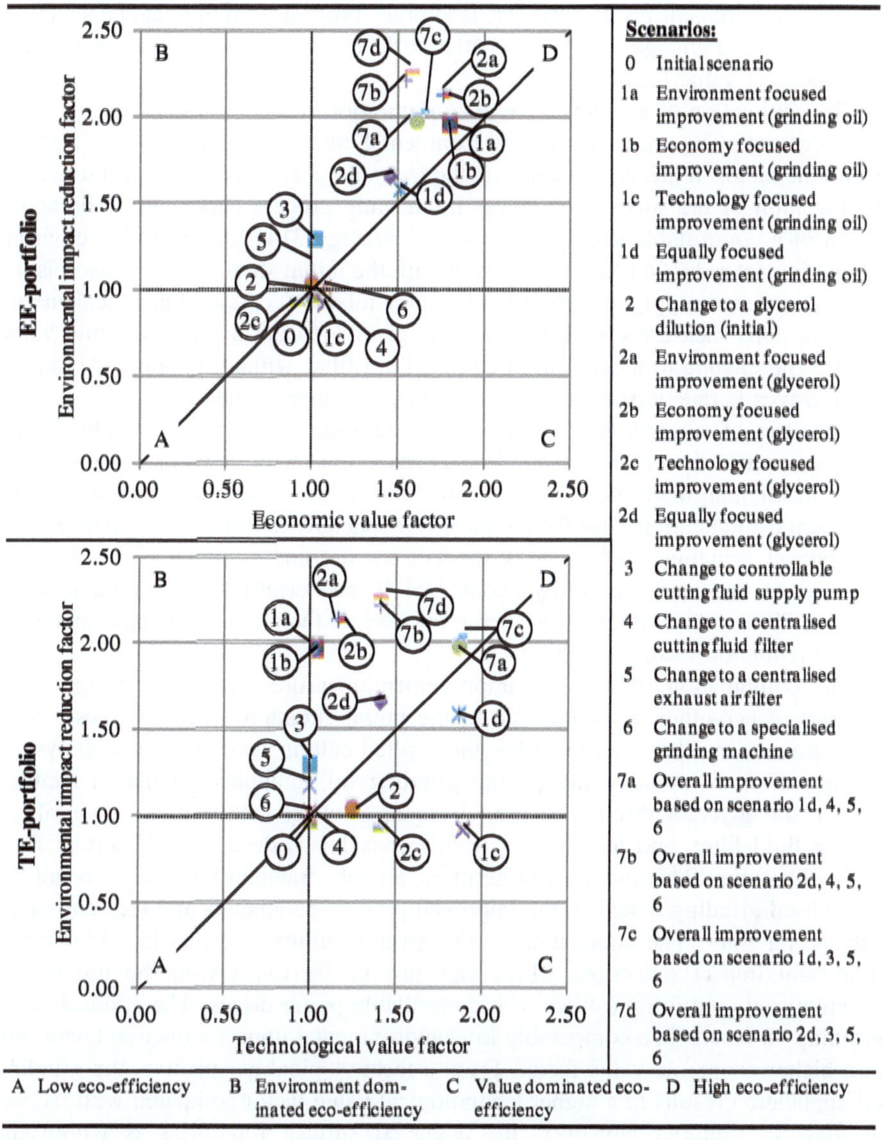

Fig. 5.5 Integrated evaluation of the external grinding process and system

The environmental impact reduction factor and the economic value factor are lower compared with strategy (1a and b). However, the technological value factor is only marginally lower. The change from the grinding oil application to the glycerol dilution (strategy 2) results in a marginal increase of the environmental impact reduction factor, while the economic value remains at the same level. The reason is a lower carbon dioxide impact of the glycerol dilution compared with the grinding oil. The procurement price, however, is similar. From a technological perspective, the cutting fluid change results in a better surface roughness and leads to a higher technological value factor.

The application of grinding system configuration-related measures results in an improvement with regard to the environmental impact reduction factor and the economic factor. It has no influence on the technological value factor, as there is no modification of the surface roughness influencing process parameters. The application of a controllable supply pump drive (strategy 3) results in a high environmental impact reduction factor compared with the initial strategy. The reason is the demand-oriented supply of cutting fluid and the related decrease of electrical energy demand. This measure also leads to a marginal increase in the economic value factor. The change to a centralised cutting fluid filter without controllable supply pump drives (strategy 5) leads only to a marginal improvement.

The use of a centralised exhaust air filtration (strategy 5) results in an improvement of the environmental impact reduction factor as well. The economic value factor marginally decreases, because the reduced expenses for energy procurement cannot compensate for the increased machine-hour rate. The change from a universal grinding machine to a specialised machine (strategy 6) results in a decrease of the overall costs by about 7.04 % and electrical energy demand by about 3.77 %. Accordingly, the economic value factor and the environmental impact reduction factor increase.

The combination of multiple improvement measures results in a significant increase of eco-efficiency compared to the initial situation. The strategies (7a–d) differ from each other with regard to the applied cutting fluid and grinding system configuration. In strategy (a)/(c), the grinding oil is applied, while in strategy (b)/(d), the glycerol dilution is used instead. Strategy (a)/(b) uses a centralised cutting fluid filter, and the strategy (c)/(d) considers a decentralised cutting fluid filter with controllable pumps. Furthermore, all sub-strategies take into account the specialised grinding machine, the improved process parameters and the centralised exhaust air filter. The application of the glycerol dilution results in a high environmental impact reduction, which can also be increased with the use of the decentralised cutting fluid filter with controllable pump drives. The application of grinding oil results in a comparably low environmental impact reduction factor, but in a higher economic value factor. From a technological perspective, the grinding oil application results in a higher technological value factor compared with the use of glycerol dilution. However, the achieved surface roughness is within the boundary conditions and significantly better compared with the initial strategy.

Step 9: Realisation

Based on the results of the integrated evaluation from step 8, the following measures should or could be applied to realise an improved grinding process and system in compliance with the lean and green requirements. The grinding process parameters should be improved in accordance with the results of strategy (1d). The application of a decentralised cutting fluid filter CFF 100 should be implemented in connection with controllable cutting fluid supply pumps. The application of a centralised exhaust air filtration unit could be taken into consideration instead of a decentralised one. The advantages of the decentralised unit, however, are its flexibility and changeability. As far as the grinding machine is concerned, a specialised machine could be favoured instead of a universal one. Yet, the universal grinding machine offers a higher degree of flexibility with regard to the internal and external grinding of bearing rings. The glycerol dilution could be applied instead of the grinding oil. The use of the dilution results in a slightly lower technological value factor and economic value factor than the oil; however, the glycerol dilution results in a higher environmental impact reduction factor. The decision depends on the focus of the lean and green factory, i.e. whether the focus is rather on "lean" or "green".

5.2 Improvement of an Existing Grinding Process

5.2.1 Initial Situation

Encouraged by the company's plans to build a benchmark lean and green manufacturing plant, an existing fictitious but realistic manufacturing plant of the same company wants to achieve a higher eco-efficiency, to prove that existing manufacturing plants can be lean and green as well. To this purpose, an internal cylindrical grinding process of bearing rings was selected for improvement. The grinding process has been operating for several years, and according to the operator, the selected process parameters represent the "best" possible combination to achieve the requirements connected with surface roughness, normal forces and specific grinding energy. Due to the existing manufacturing plant layout and limited financial resources, it is not possible to change the grinding system configuration. Yet, to increase eco-efficiency, the supervisor of this process wants to compare the impact of a change in process parameters and cutting fluid.

5.2.2 Concept Application

In the following sections, the application cycle is applied to obtain the eco-efficiency of the improved grinding process and system.

Step 1: Goal and scope definition

The purpose of the assessment is to identify and evaluate strategies for the improvement of an existing internal cylindrical grinding process of bearing rings. The results are meant to be presented to the controlling and production departments. The intended use is to support decision-making with regard to the implementation of strategies that support the achievement of a higher level of eco-efficiency. The investigated product system includes the production of bearing rings. Currently, around 75,000 bearing rings are produced each year at the existing manufacturing plant in Germany. The function of the investigated product system is the manufacturing of internal cylindrical ground bearing rings made of DIN 1.3505 featuring a hardness of circa 756 HV 5. The functional unit of the investigation is the production of one bearing featuring a workpiece surface roughness of $R_a \leq 0.25$ µm, a normal force of $F_n \leq 9.00$ N/mm and a specific grinding energy required for the material removal of a volume element of the bearing ring of $e_c \leq 45.00$ J/mm³.

Within the existing manufacturing plant, a gate-to-gate perspective represents the system boundary. Figure 5.1 shows the grinding process and system boundary and includes the upstream chains for material extraction and production of the used energy and material flows (auxiliary goods) as well as the end of life of these flows. Excluded are the upstream and downstream chains of capital goods (machinery and equipment), auxiliary resources (hydraulic fluid, etc.) as well as the transport of material and energy. No allocation to adjacent systems was made.

The value of the product system is represented by the workpiece surface roughness (technological value) and the costs (economic value). The equivalent carbon dioxide emissions are used for the representation of the environmental impact. The emissions are measured using the life cycle assessment method, due to the application of the impact category global-warming potential. Hence, the eco-efficiency can be represented either by the ratio of the technological value or by the economic value together with the environmental impact. Due to the intended use and the intended audience, both calculation methods are selected.

Step 2: Description and configuration

The grinding system consists of a grinding machine, a decentralised cutting fluid filter and a decentralised exhaust air filter. The grinding machine (type GM I120) is designed solely for internal cylindrical grinding operations. A cutting fluid filter (CFF 20) is connected and controlled by the grinding machine with a tank volume of 120 l and an exhaust air filtration unit (type EAF 1200). Both filtration units are decentralised and equipped with uncontrolled drives. A filter bag is used within the cutting fluid filter. It is assumed that the economic and environmental influence of the filtration material is negligible and thus will not be considered. Mineral oil-based grinding oil is used as a cutting fluid featuring a kinematic viscosity of 12 mm²/s. The applied cubic boron nitride grinding wheel (specification B126 T 43 V385) has an external diameter of 40 mm, a width of 15 mm and features a straight profile. The abrasive segment is vitrified bonded and coated on a steel wheel hub. The workpiece (made of DIN 1.3505) is hardened to a hardness of 756 HV 5 and features a ring-like shape. A specific material volume of 50 mm³/mm is removed per workpiece. The grinding swarf and the cutting fluid are disposed of through a hazardous waste incinerator.

To stay within the requirements of the normal force, the specific grinding energy and surface roughness, the following parameters were selected: grinding wheel speed (v_c) is 41 m/s, the workpiece speed (v_w) is 0.5 m/s, the cutting depth (a_e) is 2.0 µm, and the traverse dressing speed (v_{fad}) is 700 mm/min. The dressing speed (v_d) is directly related via a ratio of 0.8 to grinding wheel speed. The cutting fluid is applied with a pressure of seven bars and a volume flow of 20 l/min. Appendix A.4 presents the further numerical values used to describe the grinding process and system.

Step 3: Modelling

The internal cylindrical grinding process was modelled using physical as well as empirical relationships. As the grinding process and system have been in use for several years, it was possible to gather extensive empirical data. Workpiece surface roughness and tool wear volume were empirically modelled. Grinding process power, grinding forces and specific grinding energy were formulated using physical relationships.

Almost the same applies to the economic and environmental impact modelling. The empirical and physical modelling was conducted in accordance with the steps presented in Sect. 4.4.

Step 4: Validation

The validation was conducted using a second-order composite experimental design. The design takes into consideration the impact of the cutting depth, workpiece speed, cutting speed, traverse dressing speed and volume of material removed. Each parameter was modified in three steps, resulting into 52 experimental trials. During the trials, the grinding forces (normal and tangential), grinding power, surface roughness and tool wear were measured. The details of the experimental design are presented in Table 5.5.

The results of the experimental design are presented in Fig. 5.6. In the upper part, the figure shows the comparison between the observed and physically modelled normal force. Both charts are plotted according to varying process parameters. The lower chart presents the observed workpiece surface roughness and the empirically measured results.

A first analysis indicates a good match between the observed and calculated values. The surface roughness in particular shows a very good level of accuracy. In the case of the normal force, a medium level of accuracy was measured. In particular, high deviations between observed and calculated data occur at a low cutting speed and a high cutting depth. Modifications of the physical formulation for the normal force lead to minor improvements. Overall, the results coincide with the relationships presented in Sects. 4.4.2.2 and 4.4.2.3. For a detailed analysis, the coefficient of determination (R^2) and the relative absolute error (E) can be used.

Table 5.5 Parameters of the internal grinding experimental design

Process parameters	Minimum	Mean	Maximum
Cutting depth, a_e [μm]	1.00	3.00	5.00
Workpiece speed, v_w [m/s]	0.50	0.75	1.00
Cutting speed, v_c [m/s]	40.00	60.00	80.00
Spec. volume removed, V'_w [mm³/min]	30.00	165.00	300.00
Traverse dressing speed, v_{fad} [mm/min]	127.30	700.2	1273.00

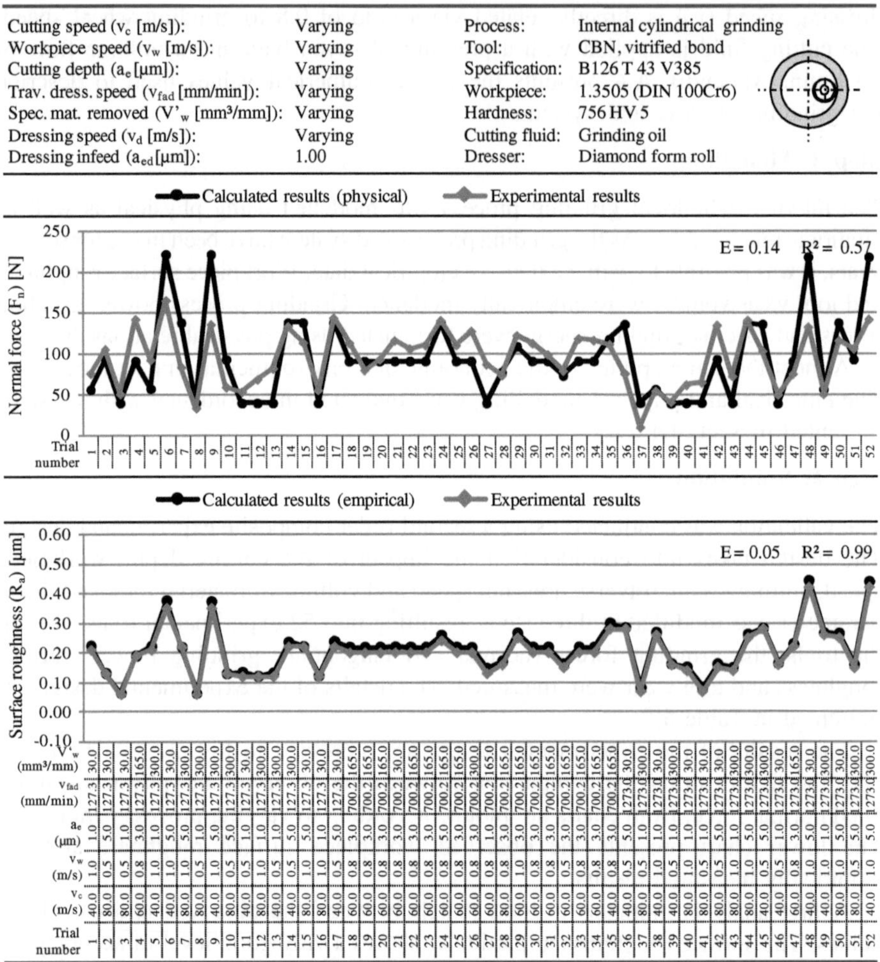

Fig. 5.6 Validation of the technological indicators

Surface roughness showed a high level of accuracy, due to a low *E*-value and a high R^2-value. In the case of the normal force, the *E*-value was very low as well, which indicates a very good match. Yet, the R^2-value shows a medium accuracy. However, if the deviations at a high cutting depth and a low cutting speed were to be minimised, then the R^2-value would increase. The reason is that some minor deviations already have a high impact. Therefore, it is necessary to take into account the *E*-value as well.

The evaluations of the predictive accuracy for the other measured technological indicators are presented in Table 5.6. The table shows, in addition to the R^2-value and *E*-value, the selected model type. The empirical models for workpiece surface roughness and tool wear volume indicate a good match between observed and

Table 5.6 Evaluation of predictive accuracy for the technological indicators

Technological indicator	Model type	Coefficient of determination (R^2) [−]	Relative absolute error (E) [−]
Tangential force (F_t)	Physical	0.734	0.285
Normal force (F_n)	Physical	0.571	0.156
Grinding power (P_c)	Physical	0.533	0.229
Workpiece roughness (R_a)	Empirical	0.994	0.046
Tool wear volume (V_t)	Empirical	0.782	0.120

calculated values. As far as the physical models are concerned, a medium fit can be determined for the normal force and the grinding power. The E-values are low, yet the R^2-value indicate a medium level of accuracy. The reason is, once again, the influence of the deviation between observed and calculated values at a low cutting speed and a high cutting depth. The tangential forces show a good level of accuracy, due to a lower E-value and a higher R^2-value in comparison with the normal force and grinding power.

Overall, it could be asserted that the physical and empirical models are capable of describing the technological indicators of the internal cylindrical grinding process with sufficient accuracy, and as a result, the models can be considered as validated.

Step 5: Single evaluation

The single indicator evaluations of the environmental and economic impact are presented in Fig. 5.7 The chart on the right shows the economic impact, while the

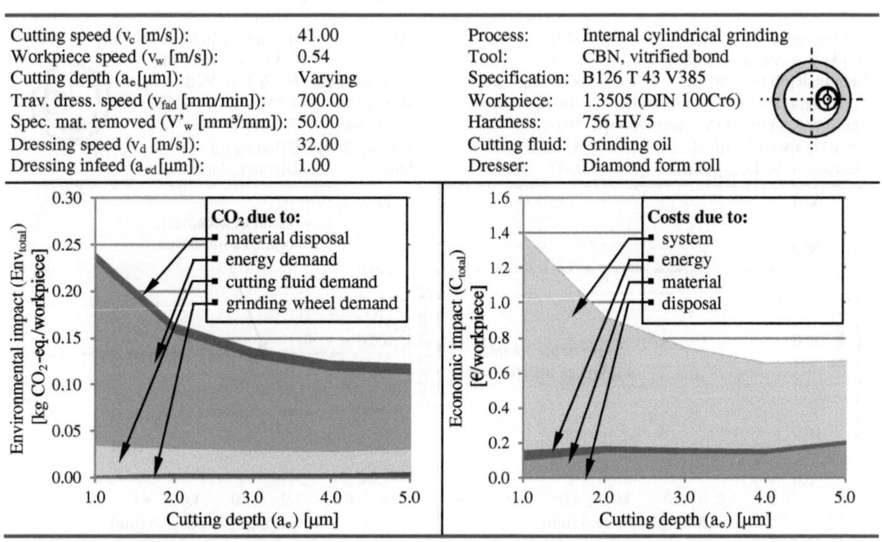

Fig. 5.7 Economic and environmental evaluation of the internal grinding process

chart on the left shows the environmental impact. Both charts measure environmental impact and economic impact against an increasing cutting depth.

The environmental impact is dominated by the impact due to the electrical energy demand, followed by the cutting fluid, the disposed material and the grinding wheel demand. The impact due to the energy demand decreases significantly with increasing cutting depth, due to the dependence of the material and energy demand with on process time. The cutting fluid demand shows a slight decrease as well. However, the demand is majorly dependent on the constant drag-out via workpiece and chip wetting. Only the cutting fluid drag-out via the exhaust air filter decreases. The wear of the grinding wheel increases with increasing cutting depth, resulting in a slight increase of the related environmental impact. In case of the economic impact, the expenses for the system are the main contributor, followed by the material costs, energy costs and disposal costs. While the system and energy costs decrease with increasing cutting depth, the material expenses increase, due to the aforementioned relations.

The existing process parameter selection features a balance concerning the consideration of workpiece interference due to the specific grinding energy and the normal force. This situation is presented in Fig. 5.8. The figure shows the development of the specific grinding energy (left side) and the normal force (right side) measured against an increasing cutting depth. The increase of the cutting depth leads to an increase of the normal force and of the specific grinding energy.

The existing grinding process is currently operated with a cutting depth of 2 μm. The figures indicate that given the existing process parameters, an increase in cutting depth would result in a level of workpiece interference that would go beyond the acceptable limits. In order to improve the grinding process, it is

Cutting speed (v_c [m/s]):	41.00	Process:	Internal cylindrical grinding
Workpiece speed (v_w [m/s]):	0.54	Tool:	CBN, vitrified bond
Cutting depth (a_e [μm]):	Varying	Specification:	B126 T 43 V385
Trav. dress. speed (v_{fad} [mm/min]):	700.00	Workpiece:	1.3505 (DIN 100Cr6)
Spec. mat. removed (V'_w [mm³/mm]):	50.00	Hardness:	756 HV 5
Dressing speed (v_d [m/s]):	32.00	Cutting fluid:	Grinding oil
Dressing infeed (a_{ed} [μm]):	1.00	Dresser:	Diamond form roll

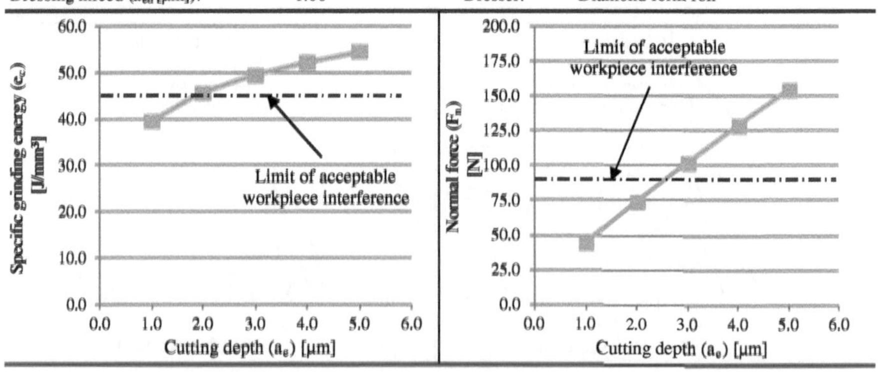

Fig. 5.8 Technological evaluation of the internal grinding process

necessary to identify process parameter combinations, which align with the technological, economic and environmental requirements.

Step 6: Improvement planning

Based on the single indicator evaluation, four improvement strategies can be implemented. The strategies focus on the grinding process only, due to the limitations of a fixed grinding system configuration. Hence, process parameters, cutting fluid and a combined change are all taken into consideration.

- **Strategy 1**: Change of process parameters. The grinding process has been operational for a couple of years, and according to the machine operators, the "best" process parameters are being used. Yet, a single- and multi-objective improvement is considered feasible. Four strategies are formulated to determine the process parameters that result in an improved environmental impact (1a), economic impact (1b), technological impact (1c) or equally weighted impact (1d).
- **Strategy 2**: Change of the cutting fluid. A mineral oil-based emulsion is applied instead of the grinding oil. The emulsion has a lower procurement price and a lower environmental impact. However, the technological capabilities might be limited. A single-objective improvement identifying process parameters resulting in a low environmental (2a), economic (2b) and technological (2c) impact is taken into consideration, as well as an equally weighted improvement strategy (2d).
- **Strategy 3**: Change of conditioning intervals. In the initial strategy and in strategy 1 and 2, the grinding wheel is dressed after each workpiece. However, investigations showed that a significantly higher number of workpieces can be manufactured during the conditioning intervals. Changes in eco-efficiency were investigated in the case of four workpieces being manufactured between each conditioning interval. Two strategies are taken into consideration, one for the grinding oil (3a) and one for the mineral oil-based emulsion (3b).
- **Strategy 4**: Combined change. In strategy (4a), a combination of strategy (1d) and (3a) is implemented in the case of the grinding oil. Strategy (4b) considers the combination of strategies (2d) and (3b) in the case of the mineral oil-based emulsion.

In each improvement strategy, only the addressed parameter is changed, while the remaining parameter details are kept unchanged.

Step 7: Multi-objective improvement

The single- and multi-objective improvement results for strategies 1 and 2 are presented in Tables 5.7 and 5.8. Each table shows the process parameters (v_c, v_w, a_e and v_{fad}) for the initial strategy and the single- and multi-objective improvement foci. The specific volume of material removed is constant at $V'_w = 50\ \text{mm}^3/\text{mm}$. The single- and multi-objective improvement is conducted taking into account the constraints defined in the goal and scope definition.

The single-objective improvement pertaining to the environmental and economic focus results in equal process parameters (Tables 5.7 and 5.8). The reason is to be

Table 5.7 Parameters resulting in an internal grinding process improvement (application of grinding oil)

Grinding process parameters	Initial	Improvement focus			
		Environmental	Economic	Technological	Equal
Grinding speed, v_c [m/s]	41.00	80.00	80.00	80.00	80.00
Workpiece speed, v_w[m/s]	0.5000	0.8000	0.8000	0.4000	0.4163
Cutting depth, a_e [μm]	2.0000	2.9000	2.9000	1.0000	4.4950
Traverse dressing speed, v_{fad}[mm/min]	700.00	1200.00	1200.00	100.00	100.00

Table 5.8 Parameters resulting in an internal grinding process improvement (application of mineral oil-based emulsion)

Grinding process parameters	Initial	Improvement focus			
		Environmental	Economic	Technological	Equal
Grinding speed, v_c [m/s]	41.00	80.00	80.00	80.00	80.00
Workpiece speed, v_w [m/s]	0.5000	0.8343	0.8343	0.4000	0.4104
Cutting depth, a_e [μm]	2.0000	5.0000	5.0000	1.0000	5.0000
Traverse dressing speed, v_{fad} [mm/min]	700.00	100.00	100.00	100.00	100.00

found in the shared data basis with regard to time, cutting fluid, electrical energy and grinding wheel demand. In the case of grinding oil application (Table 5.7), the technological single-objective improvement also results in a marginal improvement of the economic and environmental impact. Due to an increase of cutting depth, the process time is reduced by 6 %. As far as mineral oil-based emulsion is concerned, the technological focus results in a slight increase of the cutting time (by 1 %) in comparison with the initial strategy.

The multi-objective improvement of both cutting fluids leads to significant differences. The application of the grinding oil results in a better surface roughness (by 56 %), in comparison with the use of mineral oil-based emulsion. However, the application of the emulsion results in a significantly lower environmental impact (by 34 %) and lower costs (by 10 %) than the use of oil.

Step 8: Integrated evaluation

The results of the integrated evaluation are presented in Fig. 5.9. The upper diagram shows the eco-efficiency determined by the economic value factor and the environmental impact reduction factor (EE-portfolio). The lower diagram presents the ratio between the technological value factor and the environmental impact reduction factor (TE-portfolio). The results of the 14 strategies are plotted in both diagrams. The initial strategy is in the centre of the diagrams and indicates the differentiation of the four portfolio areas (A, B, C and D).

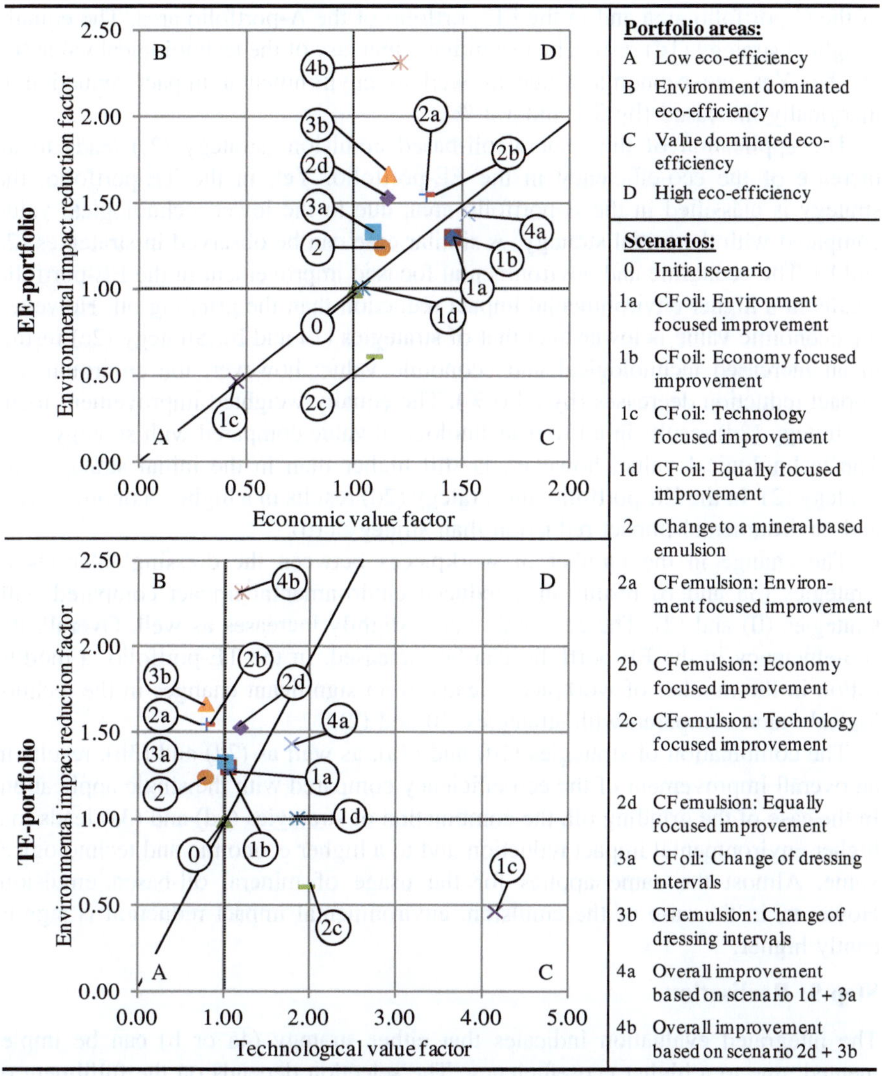

Fig. 5.9 Integrated evaluation of the internal grinding process and system

The first strategy focuses on the improvement of the process parameters when using grinding oil. The results presented in Table 5.7 already indicated that the environmental (1a) and economic improvement (1b) focus lead to the same eco-efficiency level. In the EE- and in the TE-portfolio, the environmental impact is reduced and the economic and technological value is increased compared with the initial strategy. The strategy (1c) leads to a substantial increase of the technological value (by 75.8 %) combined with a major decrease in environmental impact reduction and economic value. Hence, the strategy is classified in the TE-portfolio of the C-portfolio area and in the EE-portfolio of the A-portfolio area. The equally weighted strategy (1d) results in a significant increase of the technological value (by 46 %). Yet, the economic value as well as environmental impact reduction is marginally increased (by 3.7 and 1.2 %).

The application of the mineral oil-based emulsion (strategy (2)) leads to an increase of the eco-efficiency in the EE-portfolio. Yet, in the TE-portfolio, the strategy is classified in the B-portfolio area, due to the lower technological value compared with the initial strategy. A similar case can be observed in strategies (2a and b). The economic and environmental focused improvement in the EE-portfolio results in a higher environmental impact reduction than the grinding oil. However, the economic value is lower than that of strategies (1a and b). Strategy (2c) results in an increased technological and economic value; however, the environmental impact reduction decreases (by 64.6 %). The equally weighted improvement focus of strategy (2d) results in a lower technological value compared with strategy (1d). The technological value, however, is still higher than in the initial strategy and strategy (2). In the EE-portfolio, the strategy (2d) results in a higher economic value and environmental impact reduction than strategy (1d).

The change in the number of workpieces between the dressing intervals in strategies (3a and b) results in a reduced environmental impact compared with strategies (0) and (2). The economic value slightly increases as well. Overall, the eco-efficiency in the EE-portfolio can be increased. In the TE-portfolio, a modification in the number of workpieces leads to no significant changes in the technological value compared with strategies (0) and (2).

The combination of strategies (1d) and (3a), as well as (2d) and (3b), results in an overall improvement of the eco-efficiency compared with the single application. In the case of the grinding oil, the combination of strategies (1d) and (3a) leads to a higher environmental impact reduction and to a higher economic and technological value. Almost the same applies for the usage of mineral oil-based emulsion. However, in the case of the emulsion, environmental impact reduction is significantly higher.

Step 9: Realisation

The integrated evaluation indicates that either strategy (4a or b) can be implemented, due to a higher eco-efficiency. The selection depends on the fulfilment of the lean and green requirements. If environmental impact is the focus rather than the economic value, then strategy (4b) should be selected. If a balance between the economic and environmental improvement focus is favoured, then strategy (4a) should be selected.

Chapter 6
Concept Evaluation and Outlook

Abstract This chapter evaluates the proposed concept and presents an outlook on subsequent research extending the scope of this investigation.

6.1 Concept Evaluation

This investigation proposed and developed a concept to describe, model, evaluate and improve the eco-efficiency of grinding processes and systems. The intended application focus of the concept is to provide decision support during the planning of new or the improvement of existing grinding processes and systems. The concept is based on the state of current research presented in Chap. 3. It was developed based on the research needs that have been outlined in Sect. 3.2.3. A comprehensive concept is introduced to reduce the environmental impact while increasing the economic value in the case of grinding processes and systems based on the exploitation of the available improvement potential.

In comparison with existing approaches, this concept allows the evaluation of the influence of different grinding process and system parameters and configurations on eco-efficiency. The concept provides assistance for decision-making when planning and operating new and existing grinding processes and systems. Various combinations of grinding process parameters and system configurations can be evaluated based on eco-efficiency in order to use the most suitable ones. Furthermore, a multiple-objective improvement taking into account technological, economic and environmental impact can be conducted to implement a Pareto-optimal process parameter configuration.

The foundation and innovation of the proposed concept is the empirical and physical modelling of technological, economic and environmental impact indicators based on the relevant energy and resource flows of the grinding process and system. In addition to energy and resource flows, the physical modelling includes the formulation of grinding forces and power, workpiece surface roughness and tool wear. The formulations allow the inclusion of different grinding wheel

M. Winter, *Eco-efficiency of Grinding Processes and Systems*,
Sustainable Production, Life Cycle Engineering and Management,
DOI 10.1007/978-3-319-25205-6_6

specifications (e.g. grain type, grain size and grain concentration), workpiece materials and cutting fluid types as well as conditioning conditions and grinding processes (e.g. internal cylindrical, external cylindrical and surface grinding). The concept further includes a single and integrated evaluation as well as improvement of different grinding process and system strategies. In contrast with existing approaches, the proposed concept considers the peculiarities and relationships of and between the grinding process and system to enable a holistic evaluation. An application cycle is provided in order to transparently describe, model, evaluate and improve existing and new grinding processes and systems. The cycle allows a structured and case-dependent solving approach for various grinding process parameters and system configurations while considering the specific and general conditions during the manufacturing of ground workpieces.

Accordingly, the proposed and implemented concept presents a consistent and comprehensive approach supporting the description, modelling, evaluation and improvement of grinding processes and systems while exploiting the available improvement potential. To ensure a smooth application of the proposed concept, it is necessary to consider the following critical aspects:

- The proposed and implemented concept represents a comprehensive consideration of different energy and resource flows as well as the connected environmental and economic impact. The flows are related to each other through models representing the reality. The selected models represent the relevant influencing indicators and impact; however, not all of them could be taken into account. In-depth analyses regarding the environmental impact modelling of production, usage/maintenance and disposal of the grinding machine, cutting fluid filtration unit and exhaust air filtration unit are especially necessary. This also applies for the consideration of more complex grinding processes and profiles. A possible option would be the breakdown of a complex grinding process into single process steps with varying grinding process parameters and cutting widths. These procedures would allow the modelling of grinding processes and systems in a detailed manner, resulting in increased accuracy in the measured eco-efficiency.
- When implementing the application, several different variables need to be taken into account in order to describe the grinding wheel, workpiece, cutting fluid, grinding and conditioning process, machine tool and supporting units. The data presented in the appendix give an indication of the needed level of detail and can be used to describe a grinding process and system. To ensure a valid application of the evaluation and improvement approach, it is necessary to continuously update and to include new or changing variables. A high level of detail and the consideration of all necessary variables increase accuracy in the evaluation.
- The developed concept was realised using standard PC software products (e.g. MS Excel®, IBM SPSS®, Matlab®). The used software allows a fast and expedient evaluation of a user-specific grinding process and system featuring varying parameters and configurations without the need for specific software and programming knowledge. When implementing the approach, an exchange

of data sets between these software tools is necessary. This exchange, however, may result in errors if the user does not pay attention to the steps of the application cycle or makes mistakes when using the software/approach. The transfer into a single software solution might reduce the errors and increase concept application efficiency. However, this measure would undermine the advantage of applying standard software products.

- The modelling of the grinding process' technological indicators (e.g. grinding forces, workpiece surface roughness, grinding wheel wear, cutting fluid drag-out) is based on the case-dependent application of empirical or physical formulations. The application of physical formulations is favoured over the application of empirical ones. Yet, in some cases, the application of physical formulations leads to insufficient accuracy in predicting outcomes. Consequently, a statistically verified empirical formulation needs to be derived based on observed data. In some cases, this procedure may not be possible, thus causing a lower level of accuracy in the measured eco-efficiency. To overcome this drawback, observed data for empirical modelling must be used instead, if it is not possible to improve the predictive accuracy of existing physical formulations. However, due to the equation-based formulation of the technological impact indicators, it is possible to replace the existing formulation with more accurate ones.

6.2 Outlook

With the proposed and implemented concept, used for the evaluation and improvement of the eco-efficiency of grinding processes and systems, the research demand presented in Chap. 3 could be substantially improved. An outlook for subsequent research, based on the progress and results of this investigation, is briefly outlined in the following paragraphs. It suggests future research path, which would extend functionalities to support a further embedment.

- **General**. This investigation focuses on the specific requirements and peculiarities of the grinding process and system as a result of their major influence on the technological, economic and environmental indicators. However, the approach can be extended to physical and empirical modelling formulations representing further manufacturing processes. It is worth developing further concepts to determine, evaluate and improve the eco-efficiency of machining processes with geometrically defined cutting edges as well as forming, generative or joining processes. In addition to further manufacturing, it could be helpful to take into account other elements connected with the manufacturing process and system, such as washing machines, chip and swarf processing. This enhancement would allow an extension of the investigation to the next hierarchical level, the manufacturing system. A further option could be the coupling of the proposed concept with the technical building services described by

Thiede (2012, pp. 1ff.). This measure would allow the determination, evaluation and improvement of eco-efficiency from the chip formation level up to the factory level with one single approach.

- **Description**. The description of grinding process and system properties and parameters within this approach is based on a static database. In addition to the extension of the database with further properties, a link to live data from sensors can be considered for the purpose of extending the approach. The extension of the static database can rely on measurements, existing databases (e.g. life cycle inventory databases) or data sets provided by the producer of the applied components at the grinding process and system level. The linking of live data from sensors could include the real-time consideration of energy and material flows, such as the demand of electrical power or cutting fluid as well as the condition of the workpiece (e.g. roughness, hardness), grinding wheel (e.g. wear development) or grinding process (e.g. forces, acoustic emissions). By using this procedure, eco-efficiency can be determined in a quasi-real-time manner, allowing a direct feedback for process modification.
- **Modelling**. The modelling of the developed approach focuses on the consideration of technological, economic and environmental impact indicators. Further technological impact indicators are not taken into consideration in the approach and need to be calculated separately. However, it would be sensible to integrate further technological indicators (e.g. temperature, residual stress, chatter vibration) in the future using physical and empirical models, due to different requirements connected with workpiece quality. The aforementioned link with live data from sensors can also be used in the context of the modelling element. The linking would allow a live validation and modification of the used modelling formulations. It might also be advantageous to consider a coupling of the modelling element with a process simulation using the finite element method (FEM) or production system simulations to further extend the implemented multi-level approach. In addition to technological modelling, it can be helpful to associate the existing environmental impact models with LCA software solutions (e.g. GaBi®, Umberto®). By using this measure, the environmental assessment of used components could be further improved. An advantage of these measures would be an increased modelling accuracy and therefore an improved basis for the evaluation and improvement of the grinding process and system.
- **Evaluation**. The developed approach incorporates the evaluation of technological, economic and environmental impact indicators. While the technological impact indicator covers different categories (e.g. force, power, surface roughness), a single impact category is used to represent the environmental impact and the economic impact. In particular in the case of the environmental impact, for example, the consideration of further impact categories or the influence of cutting fluids selection on the occupational health would enable a more comprehensive assessment. The evaluation of the environmental impact in the developed concept is based on the demand of energy and resources. However, assuming that over several years the grinding system is also diminished in

ecological as well as in economical terms, the approach could be extended. To this purpose, the environmental impact due to the production, usage/maintenance and disposal phases of the grinding machine, the cutting fluid filtration unit and the exhaust air filtration unit could be considered.

- **Improvement**. The improvement step of the developed concept includes the application of a geometric programming algorithm to solve a single- or multi-objective improvement strategy. The application of a different or more complex single- and multi-objective improvement algorithm may result in a reduction of the problem-solving time especially for multi-objective improvement strategies. Currently, the solving takes a few seconds. However, if the complexity of the used physical and empirical formulations or the complexity of the strategy increases, the problem-solving time could increase significantly. Furthermore, it could be considered for the implementation of the described model in an automatic algorithm to achieve an automatic improvement of different measures and alternatives.

Reference

Thiede, S. (2012): Energy Efficiency in Manufacturing Systems. Dr.-Ing. Dissertation, Technische Universität Braunschweig, Springer Verlag, Berlin, Germany.

Chapter 7
Summary

Abstract This final chapter briefly summarises the proposed concept and the key findings of this investigation.

This investigation focuses on decision support during the planning of new or the improving of existing grinding processes and systems through the development of an approach to determine, evaluate and improve eco-efficiency. To this end, the concept shall further support the separation of the environmental impact from economic growth in manufacturing.

While Chap. 1 outlines the reasons for this investigation, Chap. 2 provides the study's relevant theoretical background. To this purpose, the known state of art is presented. The presentation starts with the introduction of grinding as an industrial production process. This presentation includes the necessary definition and characterisation of the grinding process' and system's hierarchical level (including chip formation, grinding process and grinding system). The subsystems and components at each level are specified in detail. The relationships between the various levels are presented to depict the complexity of the grinding process and system, after which the foundation for the modelling of technological, economic and environmental improvement indicators is presented. This section includes the presentation of modelling approaches, spatial and temporal dimensions and indicators. Furthermore, the chapter introduces the origin, definition and description of the eco-efficiency assessment approach. The last section of the chapter gives a preliminary conclusion.

Chapter 3 includes the review and comparative evaluation of research approaches in the context of the eco-efficiency evaluation of processes and systems. Approaches were identified and selected according to their contribution in achieving the defined indicators. The identified approaches have been classified on the basis of investigation level and scope and are evaluated according to previously defined comparison criteria. The result of this procedure was the derivation of research demand regarding the comprehensive description, modelling, evaluation and improvement of the eco-efficiency of grinding processes and systems.

M. Winter, *Eco-efficiency of Grinding Processes and Systems*,
Sustainable Production, Life Cycle Engineering and Management,
DOI 10.1007/978-3-319-25205-6_7

Based on the research demand, Chap. 4 presents the research concept developed to describe, model, evaluate and improve the eco-efficiency of grinding processes and systems. The concept comprises the following elements:

- **Description**. The description specifies the relevant properties and parameters characterising the investigated grinding process and system level. The parameters are clustered according to their influence on chip formation and process and system level, while the properties are distinguished in terms of their relationship with the technological, economic and environmental indicators. The element description provides the foundation for the following elements due to their classification of relevant properties and parameters.
- **Modelling**. With the element modelling, the energy and material flows as well as conditions (e.g. workpiece surface roughness) of the grinding process and system are represented in physical and empirical formulations for the first time. The modelling comprises technological (grinding forces, the surface roughness as well as the demand of power, grinding wheel and cutting fluid), economic (costs due to system, energy, material and waste) and environmental impact indicators (equivalent carbon dioxide emissions due to the demand of energy, cutting fluid, grinding wheel and filtration material). Modelling represents the key element of the proposed concept, due to the representation of flows and conditions in dependence with previously described properties and parameters in order to determine the indicators.
- **Evaluation**. Evaluation provides means to evaluate different grinding process and system strategies using single or integrated impact evaluation indicators. The single evaluation analyses the influence of process and system parameter variations on the technological, economic and environmental indicators. The integrated analysis conducts an eco-efficiency assessment according to the methodology presented in DIN EN ISO 14045.
- **Improvement**. The improvement step comprises the derivation of strategies to improve the grinding process and system. To this purpose, the formulated strategies take into account different technological and organisational improvement measures at the system level, while at the process level a single and multi-objective improvement algorithm is applied instead. In combination with the evaluation step, improvement provides a methodology to quantify the improvement potential of existing and new to be planned grinding processes and systems.
- **Application cycle**. On the basis of the previously introduced concept elements, a nine-step-based application cycle is defined. This cycle provides a useful support for the user to be able to describe, model, evaluate and improve the eco-efficiency of grinding processes and systems.

To introduce the applicability and potential of the developed concept, Chap. 5 provides examples of the application of the concept in the measuring, evaluation and improvement of the eco-efficiency of grinding processes and systems. To this purpose, two case studies are presented. The first case study focuses on the planning of a new and the second study on the improvement of an existing grinding process and system. In both case studies, grinding process parameters and system configurations can be derived resulting into a single- or multi-objective improvement. The case studies prove the applicability and potential of the proposed concept in the analysis of various grinding process and system strategies.

Chapter 6 concludes this investigation with a critical evaluation of the developed concept with regard to its contribution to the state of current research. The presented concept introduces a consistent and comprehensive approach in supporting the measuring, evaluation and improvement of new and existing grinding processes and systems while taking advantage of the available improvement potential. The chapter closes with an outlook on future research topics and opportunities to extend the scope of this investigation. The focus is on the potential of further research with regard to the description, modelling, evaluation and improvement elements.

Appendix

A.1 Regression Parameters

Regression coefficients grinding process (Sects. 4.4.2.2–4.4.2.4)

Variable	Exponent	F'_n	F'_t	R_a	V'_t
K		−1.59E+01	1.70E+00	−1.64E+00	9.89E−02
C		6.38E+01	7.89E+02	1.81E+00	7.50E−20
v_c	E_1^i	−6.05E−02	−7.81E+00	−3.82E−02	−3.90E−01
v_w	E_2^i	1.64E−01	5.92E+00	4.09E−03	3.06E+00
a_e	E_3^i	1.48E−01	4.13E+00	9.55E−03	2.33E+00
v_{fad}	E_4^i	−8.30E−02	3.32E+00	3.13E−02	5.24E+00
V'_w	E_5^i	−7.49E−03	−5.97E−01	1.99E−03	1.35E+00
v_d	E_6^i	6.23E−02	3.96E+00	2.42E−03	2.24E+00
d_{eq}	E_7^i	1.01E−01	4.18E+00	−3.82E−03	2.19E+00

Regression coefficients power demand cutting fluid filter pump (Sect. 4.4.2.5)

Variable	K	C	v_{CF}	p_{CF}	Q_{CF}
Exponent			$E_1^{P_{Mech}}$	$E_2^{P_{Mech}}$	$E_3^{P_{Mech}}$
Value	0.150	0.004	−0.018	1.054	0.805

Regression coefficients power demand spindle (Sect. 4.4.2.5)

Variable	$a_{1,i}$	$a_{2,i}$	b_i
Grinding wheel spindle	0.0001	1.4352	333.73
Workpiece spindle	4.892	8.3848	4.614
Dressing spindle	0.0026	0.1121	92.944

M. Winter, *Eco-efficiency of Grinding Processes and Systems*,
Sustainable Production, Life Cycle Engineering and Management,
DOI 10.1007/978-3-319-25205-6

Regression coefficients drag-out via workpiece wetting (Sect. 4.4.2.6)

Variable	A	B	C
Grinding oil	3.532	17.317	−0.257
Mineral oil based emulsion	1.051	5.563	−0.069

Regression coefficients drag-out via grinding swarf wetting (Sect. 4.4.2.6)

Variable	A	B	C
Grinding oil	0.921	0.137	−0.008
Mineral oil based emulsion	0.721	−0.014	0.003

Regression coefficients drag-out via filtration material wetting (Sect. 4.4.2.6)

Variable	A	B	C
Grinding oil	6.653	15.894	−0.142
Mineral oil based emulsion	7.401	9.466	−0.076

Regression coefficients drag-out via aerosols (Sect. 4.4.2.6)

Variable	Exponent	Grinding oil	Mineral oil based emulsion
K		−4.10E−02	3.85E−03
C		3.61E−04	1.42E−04
Q_{CF}	$E_{Q_{CF}}^{CF,1}$	1.54E−01	8.50E−01
v_e	$E_{v_e}^{CF,2}$	9.74E−01	1.81E+00
Q_{EA}	$E_{Q_{EA}}^{CF,3}$	1.68E−01	−1.85E−01
v_{CF}	$E_{v_{CF}}^{CF,4}$	7.88E−01	4.75E+01
σ_{CF}	$E_{\sigma_{CF}}^{CF,5}$	−1.03E−01	6.51E−01

A.2 Description and Configuration

Specification of the grinding process level

Cutting fluid (manufacturer and type)	Specific heat capacity of the cutting fluid	$c_{p,cf}$	J/(kg·K)
	Equivalent CO_2 impact to produce the cutting fluid	$CO_{2,p,cf}$	kg CO_2eq./kg
	Equivalent CO_2 impact to dispose the cutting fluid	$CO_{2,w,cf}$	kg CO_2eq./kg
	Diffusion coefficient of the cutting fluid	D	mm^2/s
	Interfacial friction coefficient of the cutting fluid	f_{cf}	–
	Cutting fluid procurement price	$p_{p,cf}$	€/kg
	Cutting fluid disposal price	$p_{w,cf}$	€/kg
	Service life	t_{cf}	month
	Tolerated fluid temperature rise	ΔT_{cf}	°C
	Cutting fluid orifice efficiency	η_{CFO}	–
	Cutting fluid water content	θ_{cf}	–
	Kinematic viscosity of the cutting fluid	ν_{cf}	mm^2/s
	Density of the cutting fluid	ρ_{cf}	kg/m^3
	Surface tension of the cutting fluid	σ_{cf}	mN/m
Process and system parameter	Cutting depth	a_e	µm
	Dressing depth	a_{ed}	µm
	Pressure cutting fluid	p_{cf}	bar
	Cutting fluid volume flow	Q_{cf}	l/min
	Exhaust air volume flow	Q_{ea}	m^3/h
	Air velocity	v_{air}	m/s
	Cutting speed	v_c	m/s
	Dressing tool speed	v_d	m/s
	Traverse dressing speed	v_{fad}	m/s
	Workpiece speed	v_w	m/s

(continued)

(continued)

	Workpiece material volume removed	V_w	mm^3
Workpiece (name/specification)	Equivalent CO_2 impact to dispose grinding swarf	$CO_{2,w,s}$	kg CO_2eq./kg
	Workpiece diameter	d_w	mm
	Hardness of the workpiece bulk material	$H_{b,w}$	MPa
	Hardness of the workpiece surface	$H_{s,w}$	MPa
	Grinding swarf disposal price	$p_{w,s}$	€/kg
	Density of the workpiece	ρ_w	kg/m^3
Grinding wheel and dresser (manufacturer and type)	Cutting width	a_p	mm
	Axial dressing overlap	a_{pd}	mm
	Effective dresser width	b_d	mm
	Equivalent CO_2 impact to produce a grinding wheel	$CO_{2,p,gw}$	kg CO_2eq./piece
	Equivalent CO_2 impact to dispose a grinding wheel	$CO_{2,w,gw}$	kg CO_2eq./kg
	Rotational dresser external diameter	d_d	mm
	Grinding wheel external diameter	d_s	mm
	Bond strength	F_0	N
	Hardness of the grain bulk material	$H_{b,g}$	MPa
	Hardness of the grain surface	$H_{s,g}$	MPa
	Grain concentration	K_g	%
	Grinding wheel price	$p_{p,gw}$	€/piece
	Grinding wheel disposal price	$p_{w,gm}$	€/piece
	Grit aspect ratio	q_m	–
	Ratio between uncut chip width and height	r	–
	Radius of the dresser	r_d	µm
	Mean grain radius	r_{mean}	µm
	Grain tip radius	r_g	µm
	Grain fracture strength	δ_t	GPa
	Grain loss factor	ν	–

Specification of the grinding system level

Grinding machine (manufacturer and type)	Surface area of evaporating cutting fluid	A_{cf}	m^2
	Compressed air demand	Q_{ca}	Nm^3/h
	Power demand (idle state)	$P_{GM,idle}$	W
	Machine hour rate	$p_{m,GM}$	€/h
Cutting fluid filter (manufacturer and type)	Type	Centralised/decentralised; Endless band/non-endless band	
	Area of used filtration material per second	A_{fm}	mm^2/s
	Eq. CO_2 impact filtration material production	$CO_{2,p,fm}$	kg CO_2eq./piece
	Eq. CO_2 impact filtration material disposal	$CO_{2,w,fm}$	kg CO_2eq./ m^2
	Total cutting fluid flow	$Q_{cf,total}$	l/min
	Power demand cooling unit	$P_{CFF,cl}$	W
	Power demand filtration pump	$P_{CFF,f}$	W
	Power demand lifting pump	$P_{CFF,l}$	W
	Power demand supply pump	$P_{CFF,s}$	W
	Machine hour rate	$p_{m,CFF}$	€/h
	Fill quantity	V_{cf}	l
	Electric pump drive efficiency	η_{el}	–
Exhaust air filter (manufacturer and type)	Type	Centralised/decentralised; Endless band/Non-endless band	
	Power demand (processing state)	$P_{EAF,proc}$	W
	Machine hour rate	$p_{m,EAF}$	€/h
Boundary condition	Eq. CO_2 impact electrical energy production	$CO_{2,p,e}$	(kg CO_2eq./ kWh)
	Specific compressor power demand for the compressed air generation	P'_{ca}	W/(Nm^3/ min)
	Price electrical energy	p_e	€/kWh
	Price labour	p_l	€/h
	Ambience air temperature	T_a	°C
	Ambience air humidity	ϕ_{air}	%

A.3 Case Study 1—External Cylindrical Grinding

Specification of the grinding process level

Cutting fluid (company A, grinding oil)	Specific heat capacity of cutting fluid	$c_{p,cf}$	J/(kg·K)	2094.64	Own data
	Equivalent CO_2 impact to produce cutting fluid	$CO_{2,p,cf}$	kg CO_2eq./kg	1.67	Stache et al. (2011), pp. 57ff.
	Equivalent CO_2 impact to dispose cutting fluid	$CO_{2,w,cf}$	kg CO_2eq./kg	2.85	Stache et al. (2011), pp. 57ff.
	Diffusion coefficient of the cutting fluid	D	mm^2/s	0.00	Own data
	Interfacial friction coefficient of the cutting fluid	f_{cf}	–	0.32	Own data
	Cutting fluid procurement price	$p_{p,cf}$	€/kg	2.60	Own data
	Cutting fluid disposal price	$p_{w,cf}$	€/kg	0.07	Own data
	Service life	t_{cf}	month	60.00	Own data
	Tolerated fluid temperature rise	ΔT_{cf}	°C	4.00	Own data
	Cutting fluid orifice efficiency	η_{CFO}	–	0.40	Own data
	Cutting fluid water content	θ_{cf}	–	0.00	Own data
	Kinematic viscosity of the cutting fluid	ν_{cf}	mm^2/s	12.00	Own data
	Density of the cutting fluid	ρ_{cf}	kg/m^3	890.00	Own data
	Surface tension of the cutting fluid	σ_{cf}	mN/m	29.95	Own data
Workpiece (bearing ring, DIN 1.3505)	Equivalent CO_2 impact to dispose grinding swarf	$CO_{2,w,s}$	kg CO_2eq./kg	2.38	Ecoinvent (2010)
	Workpiece diameter	d_w	mm	100.00	Own data
	Hardness of the workpiece bulk material	$H_{b,w}$	MPa	756.00	Own data
	Hardness of the workpiece surface	$H_{s,w}$	MPa	945.00	Own data
	Grinding swarf disposal price	$p_{w,s}$	€/kg	0.10	Own data
	Density of the workpiece	ρ_w	kg/m^3	7810.00	Own data

(continued)

(continued)

Cutting fluid (company B, glycerol dilution)	Specific heat capacity of cutting fluid	$c_{p,cf}$	J/(kg·K)	3342.61	Own data
	Equivalent CO_2 impact to produce cutting fluid	$CO_{2,p,cf}$	kg CO_2eq./kg	0.34	Stache et al. (2011), pp. 57ff.
	Equivalent CO_2 impact to dispose cutting fluid	$CO_{2,w,cf}$	kg CO_2eq./kg	0.42	Stache et al. (2011), pp. 57ff.
	Diffusion coefficient of the cutting fluid	D	mm^2/s	24.00	Own data
	Interfacial friction coefficient of the cutting fluid	f_{cf}	–	0.28	Own data
	Cutting fluid procurement price	$p_{p,cf}$	€/kg	3.10	Own data
	Cutting fluid disposal price	$p_{w,cf}$	€/kg	0.07	Own data
	Service life	t_{cf}	month	24.00	Own data
	Tolerated fluid temperature rise	ΔT_{cf}	°C	4.00	Own data
	Cutting fluid orifice efficiency	η_{CFO}	–	0.40	Own data
	Cutting fluid water content	θ_{cf}	%	60.00	Own data
	Kinematic viscosity of the cutting fluid	v_{cf}	mm^2/s	4.33	Own data
	Density of the cutting fluid	ρ_{cf}	kg/m^3	1125.00	Own data
	Surface tension of the cutting fluid	σ_{cf}	mN/m	24.79	Own data
Process and system parameter	Cutting depth	a_e	µm	varies	Own data
	Dressing depth	a_{ed}	µm	1.00	Own data
	Cutting width	a_p	mm	10.00	Own data
	Pressure cutting fluid	p_{cf}	bar	7.00	Own data
	Cutting fluid volume flow	Q_{cf}	l/min	100.00	Own data
	Exhaust air volume flow	Q_{ea}	m^3/min	15.00	Own data
	Air velocity	v_{air}	m/s	10.00	Own data
	Cutting speed	v_c	m/s	varies	Own data
	Dressing tool speed	v_d	m/s	varies	Own data
	Traverse dressing speed	v_{fad}	m/s	varies	Own data
	Workpiece speed	v_w	m/s	varies	Own data
	Workpiece volume removed	V_w	mm^3	1000.00	Own data

(continued)

(continued)

Grinding wheel and dresser (company C, B126 V VV 80)	Cutting width	a_p	mm	10.00	Own data
	Axial dressing overlap	a_{pd}	mm	1.50	Own data
	Effective dresser width	b_d	mm	0.45	Own data
	Equivalent CO_2 impact to produce a grinding wheel per dressing stroke	$CO_{2,p,gw}$	kg CO_2eq./stroke	9.79E−03	Winter et al. (2015), pp. 15ff.
	Equivalent CO_2 impact to dispose a grinding wheel	$CO_{2,w,gw}$	kg CO_2eq./kg	1.89E−03	Winter et al. (2015), pp. 15ff.
	Rotational dresser external diameter	d_d	mm	120.00	Own data
	Grinding wheel external diameter	d_s	mm	400.00	Own data
	Bond strength	F_0	N	75.00	Pelenik et al. (1964), pp. 296
	Hardness of the grain bulk material	$H_{b,g}$	GPa	45.00	Davis (1995), pp. 101ff
	Hardness of the grain surface	$H_{s,g}$	GPa	56.00	Assumption
	Grain concentration	K_g	%	25.00	Own data
	Grinding wheel price per dressing stroke	$p_{p,gw}$	€/stroke	0.31	Own data
	Grinding wheel disposal price per dressing stroke	$p_{w,gm}$	€/stroke	4.13E−03	Own data
	Grit aspect ratio	q_m	–	1.41	Heuer (1992), pp. 30
	Ratio between uncut chip width and height	r	–	8.50	Own data
	Radius of the dresser	r_d	mm	25.31	Own data
	Mean grain radius	r_{mean}	µm	57.75	Own data
	Grain tip radius	r_g	µm	13.00	Rabiey (2010), pp. 127
	Grain fracture strength	δ_t	GPa	2.34	Ichida et al. (2010), pp. 1006
	Grain loss factor	ν	–	0.75	Assumption

Specification of the grinding system level

Grinding machine (GM U40)	Surface area of evaporating cutting fluid	A_{cf}	m^2	2.00	Own data
	Compressed air demand	Q_{ca}	Nm^3/h	29.30	Own data
	Power demand (idle state)	$P_{GM,idle}$	W	3680.00	Own data
	Machine hour rate	$p_{m,GM}$	€/h	20.79	Own data
Grinding machine (GM E22)	Surface area of evaporating cutting fluid	A_{cf}	m^2	2.00	Own data
	Compressed air demand	Q_{ca}	Nm^3/h	29.30	Own data
	Power demand (idle state)	$P_{GM,idle}$	W	3680.00	Own data
	Machine hour rate	$p_{m,GM}$	€/h	20.79	Own data
Exhaust air filter (EAF 1200)	Type	Decentralised filter for one machine; application of a mechanical filtration			
	Power demand (processing state)	$P_{EAF,proc}$	W	1150.00	Own data
	Machine hour rate	$p_{m,EAF}$	€/h	0.43	Own data
Exhaust air filter (EAF 12000)	Type	Centralised filter. Ten machines à 20 m^3/min; application of endless band filtration material			
	Power demand (processing state)	$P_{EAF,proc}$	W	3940.00	Own data
	Machine hour rate	$p_{m,EAF}$	€/h	0.18	Own data
Boundary condition	Eq. CO_2 impact electrical energy production	$CO_{2,p,e}$	kg CO_2eq./kJ	1.65	UBA (2014), pp. 2
	Specific compressor power demand for the compressed air generation	P'_{ca}	W/(Nm^3/min)	8.48	Own data
	Price electrical energy	p_e	€/kWh	0.156	BDEW (2014), pp. 11
	Price labour	p_l	€/h	36.77	Schröder (2014), pp. 6
	Ambience air temperature	T_a	°C	21.00	Own data
	Workpiece handling time	t_h	s	15.00	Own data
	Tool travel time	t_t	s	5.00	Own data
	Ambience air humidity	ϕ_{air}	%	50.00	Own data
Cutting fluid filter (CFF 100)	Type	Decentralised filter for one machine, application of endless band filtration			
	Area of used filtration material per second	A_{fm}	mm^2/s	83.30	Own data
	Eq. CO_2 impact filtration material production	$CO_{2,p,fm}$	kg CO_2eq./piece	0.00	Own data
	Eq. CO_2 impact filtration material disposal	$CO_{2,w,fm}$	kg CO_2eq./m^2	0.00	Own data
	Total cutting fluid flow	$Q_{cf,total}$	l/min	100.00	Own data

(continued)

(continued)

	Power demand cooling unit	$P_{CFF,el}$	W	4280.00	Own data
	Power demand filtration pump	$P_{CFF,f}$	W	1120.00	Own data
	Power demand lifting pump	$P_{CFF,l}$	W	1100.00	Own data
	Power demand supply pump	$P_{CFF,s}$	W	8465.00	Own data
	Machine hour rate	$p_{m,CFF}$	€/h	4.23	Own data
	Fill quantity	V_{cf}	l	2000.00	Own data
	Electric pump drive efficiency	η_{el}	–	0.85	Own data
Cutting fluid filter (CFF 1000)	Type	Centralised filter for ten machines à 100 l/min; application of endless band filtration material			
	Area of used filtration material per second	A_{fm}	mm^2/s	340.00	Own data
	Eq. CO_2 impact filtration material production	$CO_{2,p,fm}$	kg CO_2eq./piece	0.00	Own data
	Eq. CO_2 impact filtration material disposal	$CO_{2,w,fm}$	kg CO_2eq./m^2	0.00	Own data
	Total cutting fluid flow	$Q_{cf,total}$	l/min	1000.00	Own data
	Power demand cooling unit	$P_{CFF,el}$	W	40,000.00	Own data
	Power demand filtration pump	$P_{CFF,f}$	W	20,000.00	Own data
	Power demand lifting pump	$P_{CFF,l}$	W	0.00	Own data
	Power demand supply pump	$P_{CFF,s}$	W	20,000.00	Own data
	Machine hour rate	$p_{m,CFF}$	€/h	0.92	Own data
	Fill quantity	V_{cf}	l	20,000.00	Own data
	Electric pump drive efficiency	η_{el}	–	0.85	Own data

A.4 Case Study 2—Internal Cylindrical Grinding

Specification of the grinding process level

Workpiece (bearing ring, DIN 1.3505)	Equivalent CO_2 impact to dispose a grinding swarf	$CO_{2,w,s}$	kg CO_2eq./kg	2.38	Ecoinvent (2010)
	Workpiece diameter	d_W	mm	100.00	Own data
	Hardness of the workpiece bulk material	$H_{b,W}$	MPa	756.00	Own data
	Hardness of the workpiece surface	$H_{s,W}$	MPa	945.00	Own data
	Grinding swarf disposal price	$p_{w,s}$	€/kg	0.10	Own data
	Density of the workpiece	ρ_W	kg/m³	7810.00	Own data
Cutting fluid (company A, grinding oil)	Specific heat capacity of cutting fluid	$c_{p,cf}$	J/(kg·K)	2094.64	Own data
	Equivalent CO_2 impact to produce cutting fluid	$CO_{2,p,cf}$	kg CO_2eq./kg	1.67	Stache et al. (2011), pp. 57ff.
	Equivalent CO_2 impact to dispose cutting fluid	$CO_{2,w,cf}$	kg CO_2eq./kg	2.85	Stache et al. (2011), pp. 57ff.
	Diffusion coefficient of the cutting fluid	D	mm²/s	0.00	Own data
	Interfacial friction coefficient of the cutting fluid	f_{cf}	–	0.32	Own data
	Cutting fluid procurement price	$p_{p,cf}$	€/kg	2.60	Own data
	Cutting fluid disposal price	$p_{w,cf}$	€/kg	0.07	Own data
	Service life	t_{cf}	month	60.00	Own data
	Tolerated fluid temperature rise	ΔT_{cf}	°C	4.00	Own data
	Cutting fluid orifice efficiency	η_{CFO}	–	0.40	Own data
	Cutting fluid water content	θ_{cf}	–	0.00	Own data
	Kinematic viscosity of the cutting fluid	ν_{cf}	mm²/s	12.00	Own data
	Density of the cutting fluid	ρ_{cf}	kg/m³	890.00	Own data
	Surface tension of the cutting fluid	σ_{cf}	mN/m	29.95	Own data

(continued)

(continued)

Grinding wheel and dresser (company C, B126 V VV 80)	Cutting width	$a_p r$	mm	10.00	Own data
	Axial dressing overlap	a_{pd}	mm	1.50	Own data
	Effective dresser width	b_d	mm	0.30	Own data
	Equivalent CO_2 impact to produce a grinding wheel per dressing stroke	$CO_{2,p,gw}$	kg CO_2eq./stroke	1.05E-03	Winter et al. (2015), pp. 15ff
	Equivalent CO_2 impact to dispose a grinding wheel	$CO_{2,w,gw}$	kg CO_2eq./kg	2.72E−04	Winter et al. (2015), pp. 15ff
	Rotational dresser external diameter	d_d	mm	120.00	Own data
	Grinding wheel external diameter	d_s	mm	40.00	Own data
	Bond strength	F_0	N	75.00	Pelenik et al. (1964), pp. 296
	Hardness of the grain bulk material	$H_{b,g}$	GPa	45.00	Davis (1995), pp. 101ff
	Hardness of the grain surface	$H_{s,g}$	GPa	56.00	Assumption
	Grain concentration	K_g	%	50.00	Own data
	Grinding wheel price per dressing stroke	$p_{p,gw}$	€/stroke	0.04	Own data
	Grinding wheel disposal price per dressing stroke	$p_{w,gm}$	€/stroke	4.03E−06	Own data
	Grit aspect ratio	q_m	–	1.41	Heuer (1992), pp. 30
	Ratio between uncut chip width and height	r	–	7.50	Own data
	Radius of the dresser	r_d	mm	25.31	Own data
	Mean grain radius	r_{mean}	μm	57.75	Own data
	Grain tip radius	r_g	μm	13.00	Rabiey (2010), pp. 127
	Grain fracture strength	δ_t	GPa	2.34	Ichida et al. (2010), pp. 1006
	Grain loss factor	v	–	0.50	Assumption

(continued)

(continued)

Cutting fluid (company B, mineral oil based dilution)	Specific heat capacity of cutting fluid	$c_{p,cf}$	J/(kg•K)	3342.61	Own data
	Equivalent CO_2 impact to produce cutting fluid	$CO_{2,p,cf}$	kg CO_2eq./kg	0.34	Stache et al. (2011), pp. 57ff
	Equivalent CO_2 impact to dispose cutting fluid	$CO_{2,w,cf}$	kg CO_2eq./kg	0.42	Stache et al. (2011), pp. 57ff
	Diffusion coefficient of the cutting fluid	D	mm^2/s	24.00	Own data
	Interfacial friction coefficient of the cutting fluid	f_{cf}	–	0.28	Own data
	Cutting fluid procurement price	$p_{p,cf}$	€/kg	0.23	Own data
	Cutting fluid disposal price	$p_{w,cf}$	€/kg	0.07	Own data
	Service life	t_{cf}	month	12.00	Own data
	Tolerated fluid temperature rise	ΔT_{cf}	°C	4.00	Own data
	Cutting fluid orifice efficiency	η_{CFO}	–	0.40	Own data
	Cutting fluid water content	θ_{cf}	%	92.50	Own data
	Kinematic viscosity of the cutting fluid	v_{cf}	mm^2/s	0.87	Own data
	Density of the cutting fluid	ρ_{cf}	kg/m^3	1010.00	Own data
	Surface tension of the cutting fluid	σ_{cf}	mN/m	31.17	Own data
Process and system parameter	Cutting depth	a_e	μm	varies	Own data
	Dressing depth	a_{ed}	μm	1.00	Own data
	Cutting width	a_p	mm	10.00	Own data
	Pressure cutting fluid	p_{cf}	bar	7.00	Own data
	Cutting fluid volume flow	Q_{cf}	l/min	20.00	Own data
	Exhaust air volume flow	Q_{ea}	m^3/min	15.00	Own data
	Air velocity	v_{air}	m/s	10.00	Own data
	Cutting speed	v_c	m/s	varies	Own data
	Dressing tool speed	v_d	m/s	varies	Own data
	Traverse dressing speed	v_{fad}	m/s	varies	Own data
	Workpiece speed	v_w	m/s	varies	Own data
	Workpiece volume removed	V_w	mm^3	500.00	Own data

Specification of the grinding system level

Grinding machine (GM I120)	Surface area of evaporating cutting fluid	A_{cf}	m^2	1.00	Own data
	Compressed air demand	Q_{ca}	Nm^3/h	29.30	Own data
	Power demand (idle state)	$P_{GM,idle}$	W	3680.00	Own data
	Machine hour rate	$p_{m,GM}$	€/h	13.52	Own data
Exhaust air filter (EAF 1200)	Type	Decentralised filter for one machine; application of a mechanical filtration			
	Power demand (processing state)	$P_{EAF,proc}$	W	1150.00	Own data
	Machine hour rate	$p_{m,EAF}$	€/h	0.43	Own data
Cutting fluid filter (CFF 20)	Type	Decentralised filter for one machine, application of endless band filtration			
	Area of used filtration material per second	A_{fm}	mm^2/s	0.00	Own data
	Eq. CO_2 impact filtration material production	$CO_{2,p,fm}$	kg CO_2eq./piece	0.00	Own data
	Eq. CO_2 impact filtration material disposal	$CO_{2,w,fm}$	kg CO_2eq./m^2	0.00	Own data
	Total cutting fluid flow	$Q_{cf,total}$	l/min	20.00	Own data
	Power demand cooling unit	$P_{CFF,cl}$	W	850.00	Own data
	Power demand filtration pump	$P_{CFF,f}$	W	850.00	Own data
	Power demand lifting pump	$P_{CFF,l}$	W	0.00	Own data
	Power demand supply pump	$P_{CFF,s}$	W	0.00	Own data
	Machine hour rate	$p_{m,CFF}$	€/h	0.98	Own data
	Fill quantity	V_{cf}	l	120	Own data
	Electric pump drive efficiency	η_{el}	–	0.85	Own data
Boundary condition	Eq. CO_2 impact electrical energy production	$CO_{2,p,e}$	(kg CO_2eq./kJ)	1.65	UBA (2014), pp. 2
	Specific compressor power demand for the compressed air generation	P'_{ca}	$W/(Nm^3/min)$	8.48	Own data
	Price electrical energy	p_e	€/kWh	0.156	BDEW (2014), pp. 11
	Price labour	p_l	€/h	36.77	Schröder (2014), pp. 6
	Ambience air temperature	T_a	°C	21.00	Own data
	Workpiece handling time	t_h	s	15.00	Own data
	Tool travel time	t_t	s	5.00	Own data
	Ambience air humidity	ϕ_{air}	%	50.00	Own data

References

BDEW (2014): Energie-Info - Industriestrompreis. Bundesverband der Energie- und Wasserwirtschaft e.V., Berlin, Germany.

Davis, J.R. (1995): ASM Specialty Handbook Tool Materials. ASM International, Ohio, USA.

Ecoinvent (2011): Database ecoinvent Data v2.2. Centre for Life Cycle Inventories, Dübendorf, Switzerland.

Heuer, W. (1992): Außenrundschleifen mit kleinen keramisch gebundenen CBN-Schleifscheiben. Dr.-Ing. Dissertation, TU Hannover, Hannover, Germany.

Ichida, Y., Fujimoto, M., Inoue, Y., Matsui, K. (2010): Development of High-Performance Vitrified Grinding Wheels using Ultrafine-Crystalline cBN Abrasive Grains. Journal of Advanced Mechanical Design Systems and Manufacturing, 4:1005-1014.

Pelenik, J., Lane, R., Shaw, M.C. (1964): Comparison of static and dynamic hardness of grinding wheels. Journal of Manufacturing Science and Engineering, 86/3:294-297.

Rabiey, M. (2010): Dry Grinding with cBN Wheels - The effect of structuring. Dr.-Ing. Dissertation, Universität Stuttgart, Jost-Jetter Verlag, Stuttgart, Germany.

Schröder, C. (2014): Industrielle Arbeitskosten im internationalen Vergleich. IW-Trends, 04/2014, Köln, Germany.

Stache, H., Winter, M., Wichmann, H., Bock, R., Öhlschläger, G., Dettmer, T., Herrmann, C., Bahadir, M., Draeger, H., Dwuletzki, H. (2011): Entwicklung eines Glycerin-basierten biozidfreien Kühlschmierstoffs für die industrielle Metallbearbeitung. DBU Abschlussbericht AZ 25369-31, Braunschweig, Germany.

UBA (Umweltbundesamt) (2014): Entwicklung der spezifischen Kohlendioxid-Emissionen des deutschen Strommix in den Jahren 1990 bis 2013. Climate Change 23/2014, Dessau-Roßlau, Germany.

Winter, M., Ibbotson, S, Kara, S., Herrmann, C. (2015): Life cycle assessment of cubic boron nitride grinding wheels. Journal of Cleaner Production, 107:707–721.

MIX
Papier aus verantwortungsvollen Quellen
Paper from responsible sources
FSC® C105338

If you have any concerns about our products,
you can contact us on
ProductSafety@springernature.com

In case Publisher is established outside the EU,
the EU authorized representative is:
Springer Nature Customer Service Center GmbH
Europaplatz 3, 69115 Heidelberg, Germany

Printed by Libri Plureos GmbH
in Hamburg, Germany